Super Smart Science Series
COLLECTION
Books 6 through 10

Anatomy & Physiology Part 2:
Body Systems

Cardiology FOR KIDS! ...and Adults Too!

Botany:
Plants, Cells & Photosynthesis

Marine Biology

Geology:
Earth Composition, Landforms, Rocks & Water

Dedicated to:
MOM AND DAD

Super Smart Science Series Combo Book : 6 through 10
ISBN #: 978-1-941775-15-8
April Chloe Terrazas, BS University of Texas at Austin
© 2015 Crazy Brainz Publishing

Visit us on the web! www.Crazy-Brainz.com

ANATOMY and Physiology PART 2: Body Systems

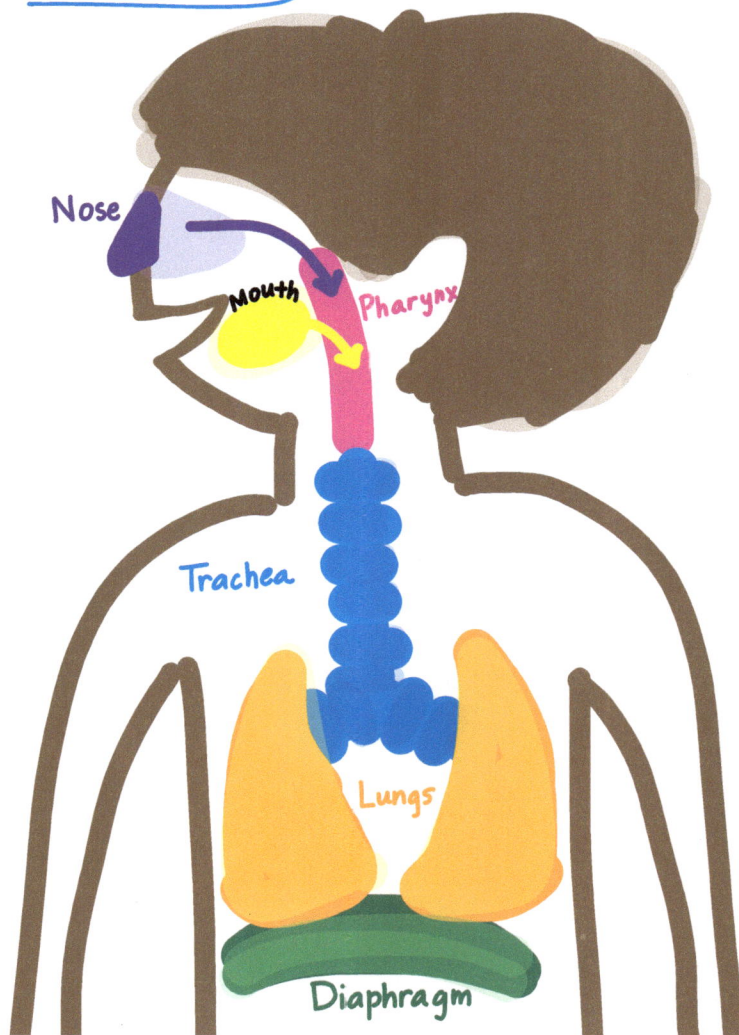

Nose

Mouth

Pharynx

Trachea

Lungs

Diaphragm

By: April Chloe Terrazas

Book 6 of the Super Smart Science Series
Ages 0-100

This SUPER AWESOME Book Belongs to:

_____

April Chloe Terrazas

Winner of the Art Competition

The Digestive Syetem

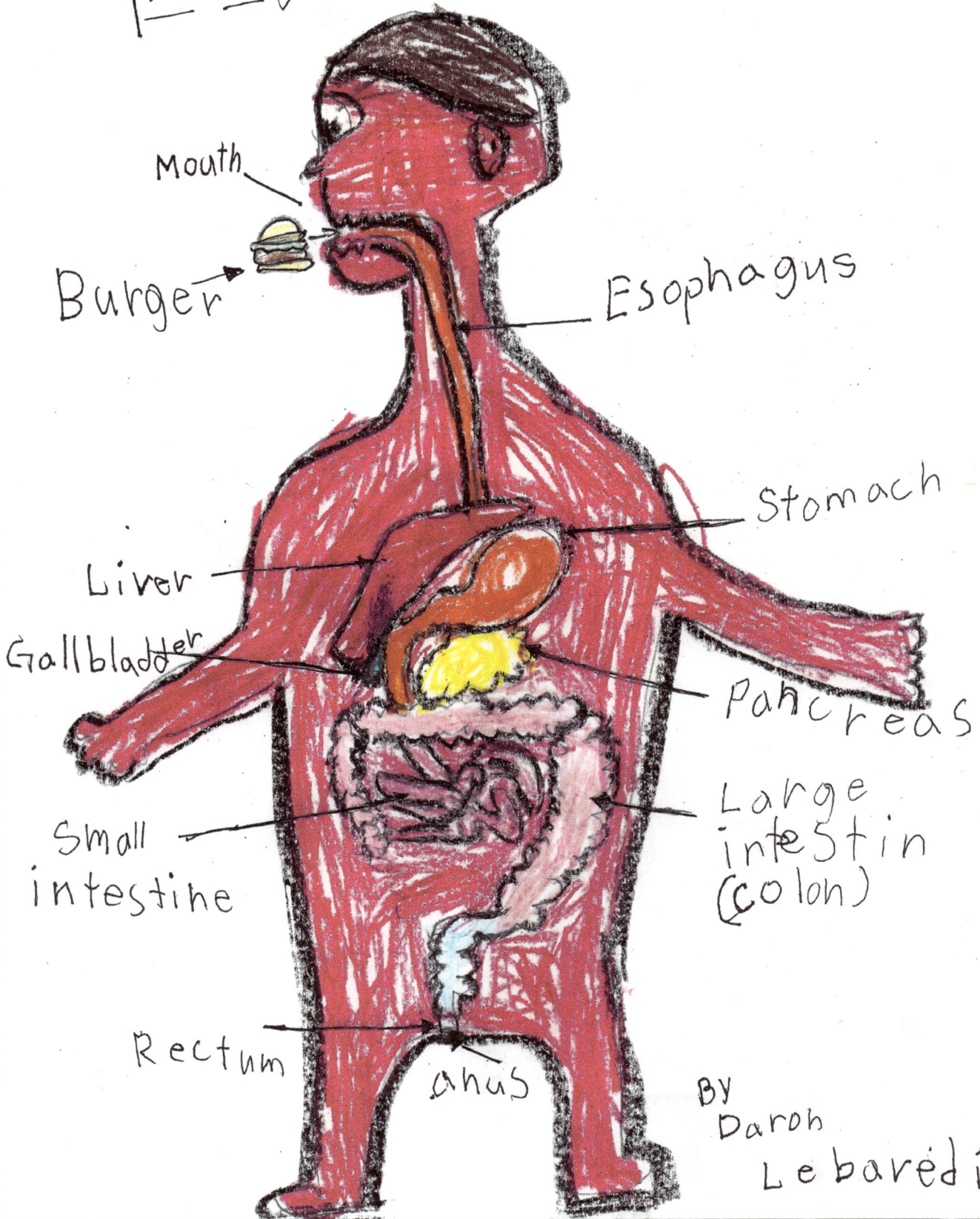

Mouth

Burger

Esophagus

Stomach

Liver

Gallbladder

Pancreas

Small intestine

Large intestin (colon)

Rectum

anus

BY Daron Lebaredian

ANATOMY and Physiology PART 2: Body Systems

By:
April Chloe Terrazas

This book is dedicated to:

Elias & Diana Urbina, whose love and support I will cherish forever.

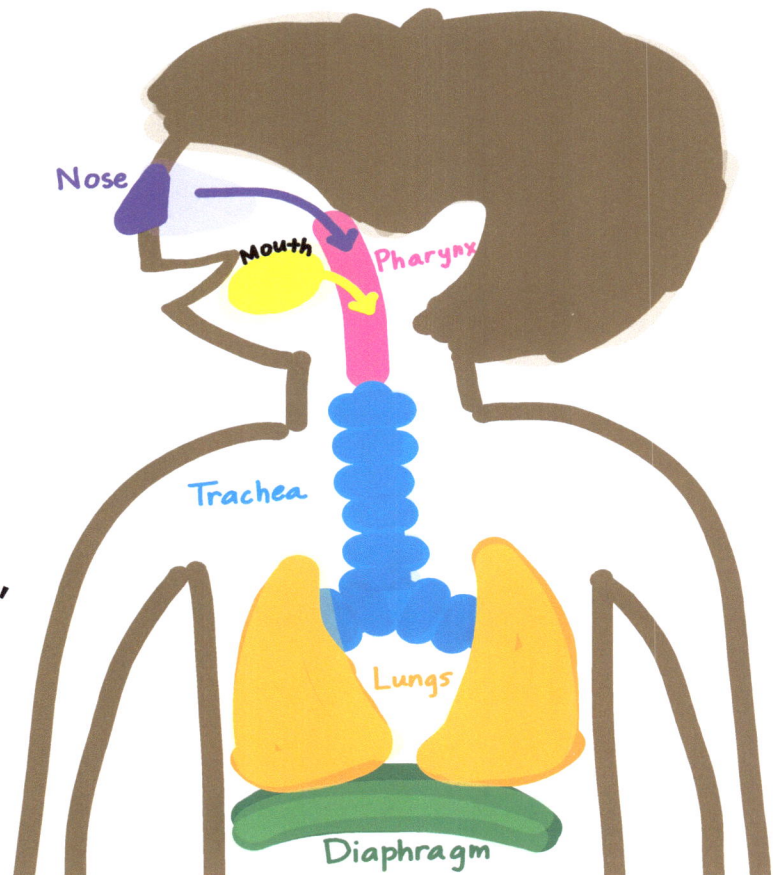

Nose

Mouth

Pharynx

Trachea

Lungs

Diaphragm

Anatomy & Physiology PART 2: Body Systems
April Chloe Terrazas, BS University of Texas at Austin.
Copyright © 2014 Crazy Brainz, LLC

Visit us on the web! www.Crazy-Brainz.com

Cover design, illustrations and text by: April Chloe Terrazas

Body systems are groups of organs that work together for your body to function properly. Right now, you are using all of your body systems!

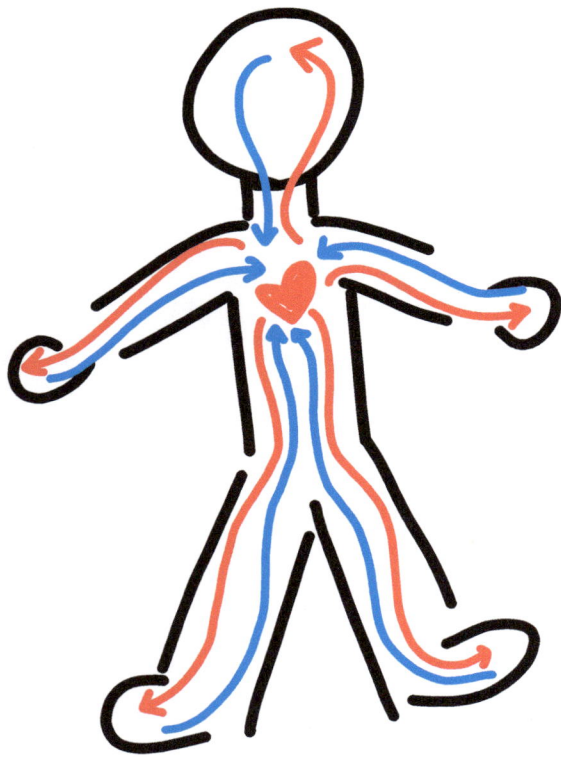

Are you ready to learn about body systems?

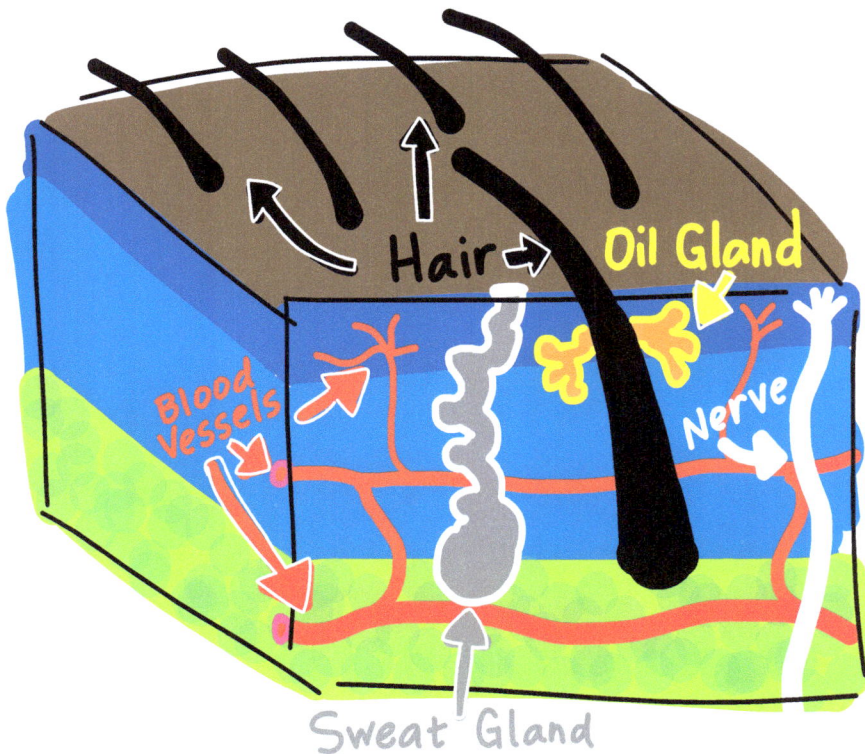

NORMAL

Hair

Oil Gland

Blood Vessels

Nerve

Sweat Gland

Circulatory System

The **circulatory system** is a network of vessels throughout your body. They are everywhere!

The **circulatory system** is like a railroad. It moves blood and chemicals all over the body.

The main station on the circulatory system railroad is the heart!

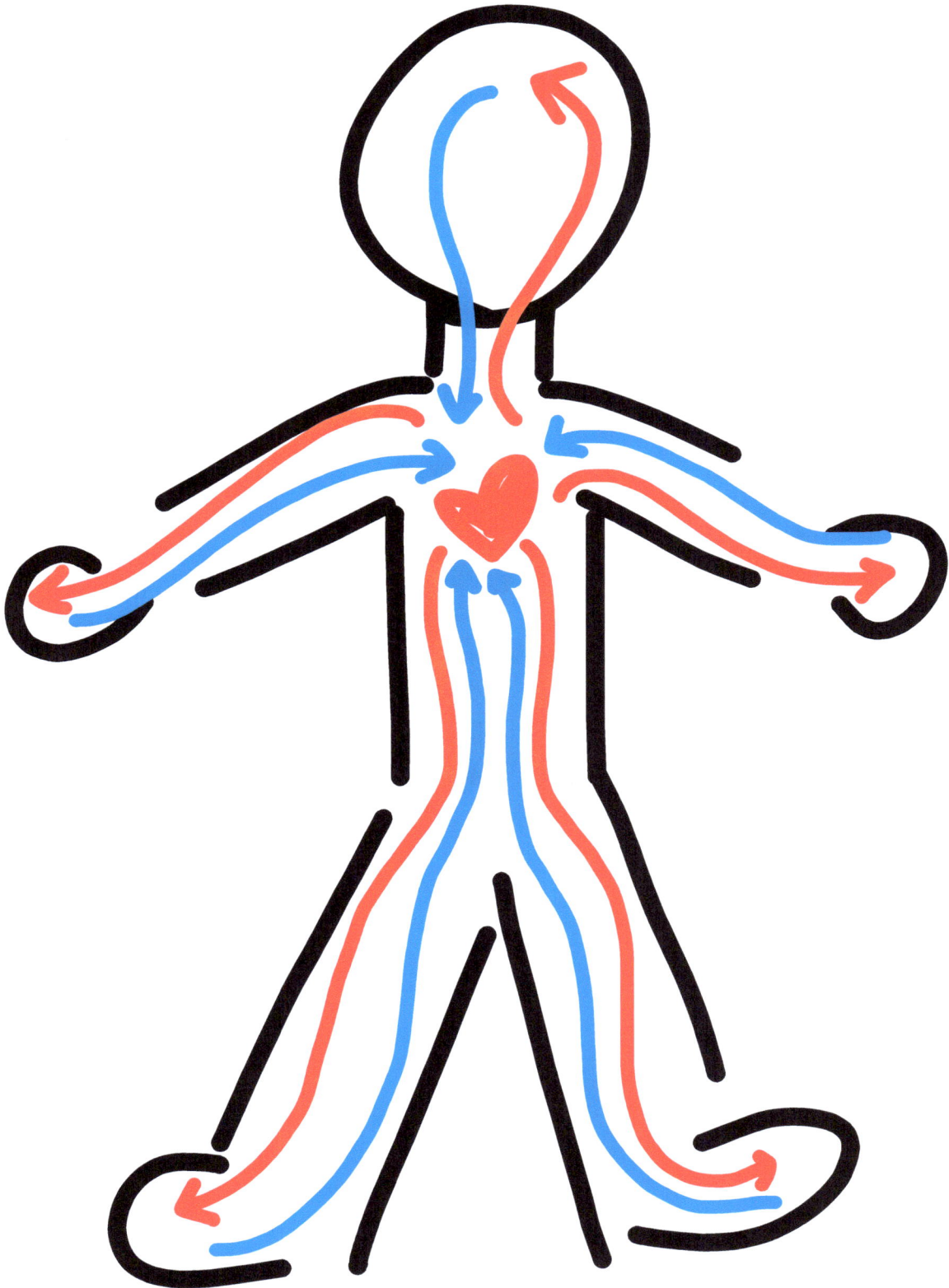

The circulatory system railroad goes to every organ in the body.

All of the cells in our body need oxygen to survive.

The circulatory system is connected with all of our cells and delivers oxygen.

Take a deep breath...
...and let it out.

Your circulatory system took oxygen from the lungs and gave back CO_2 to be exhaled.

Respiratory System

The respiratory system works with the circulatory system.

Oxygen is breathed in through the respiratory system and from there is transferred to the blood in the circulatory system.

Then the circulatory system delivers oxygen to all of your cells and comes back to the lungs to transfer CO_2 out of the body.

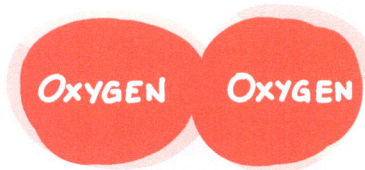

OXYGEN OXYGEN

Take a deep breath...

...let it out.

OXYGEN CARBON OXYGEN

Oxygen was transferred into the blood stream (circulatory system) and carbon dioxide was transferred out of the blood stream and breathed out through the respiratory system.

The **respiratory system** is made of the following parts:

Nose: NOZ

Mouth: MOWTH

Pharynx: FAR-INX

Trachea: TRA-KEE-UH

Lungs: LUNGZ

Diaphragm: DI-UH-FRAM

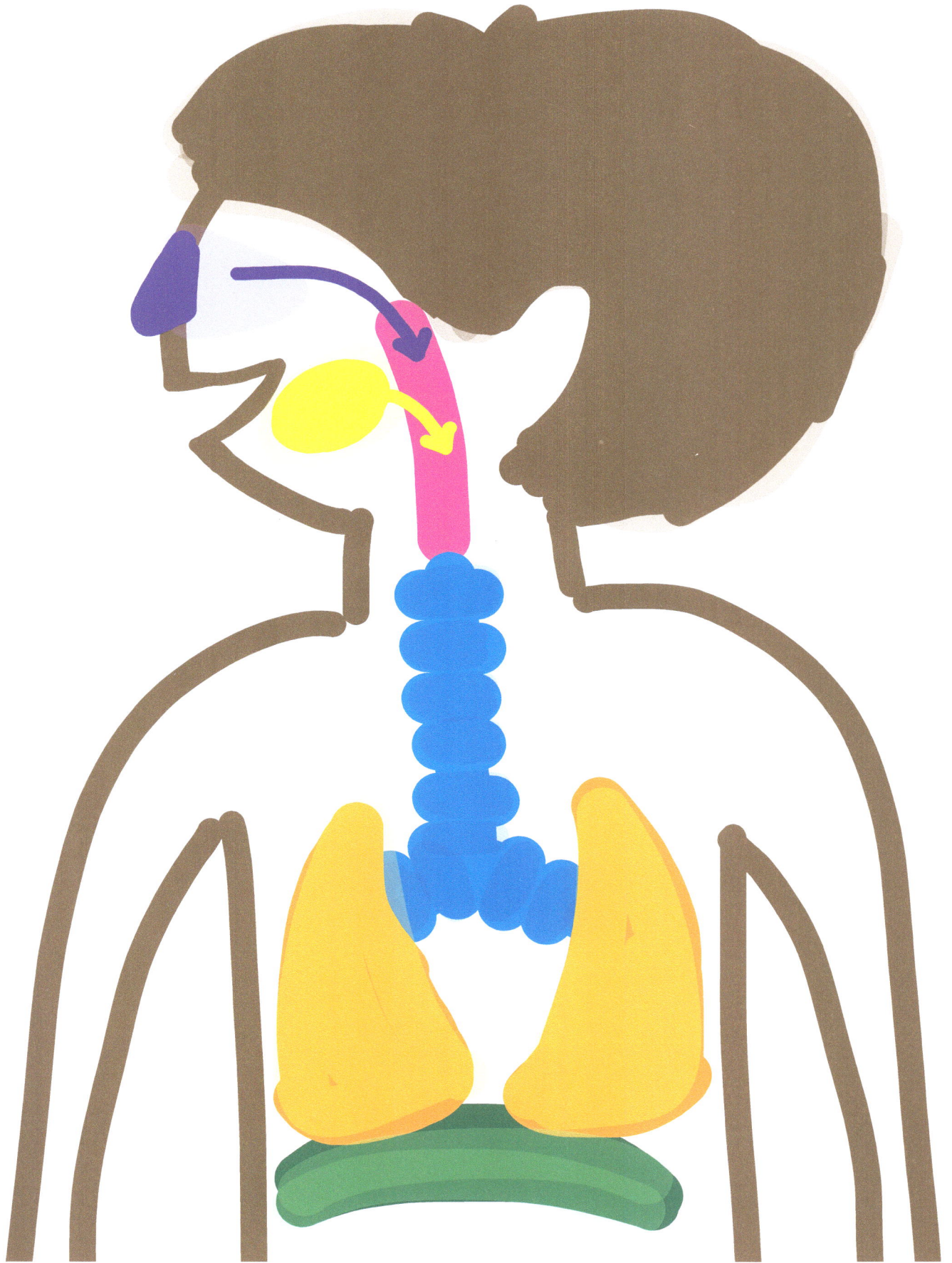

We need oxygen to survive. We especially need it to be active and play!

When you run very fast, what happens to your breathing?

Oxygen is coming in faster and CO_2 is going out faster.

Oxygen goes in the nose or mouth, through the pharynx, down the trachea, and into the lungs. Air is pushed in and out of the lungs by the diaphragm.

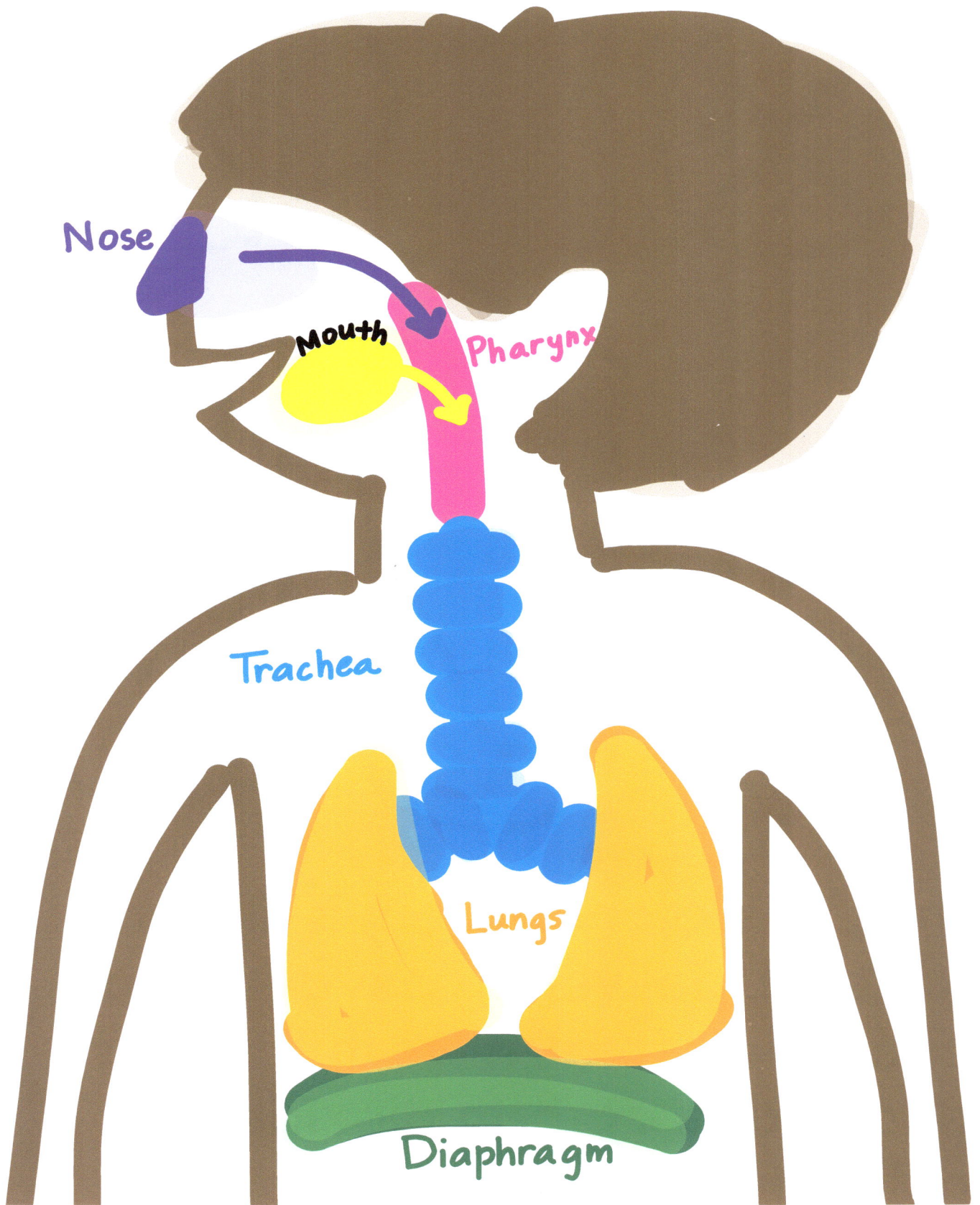

Nose

Mouth

Pharynx

Trachea

Lungs

Diaphragm

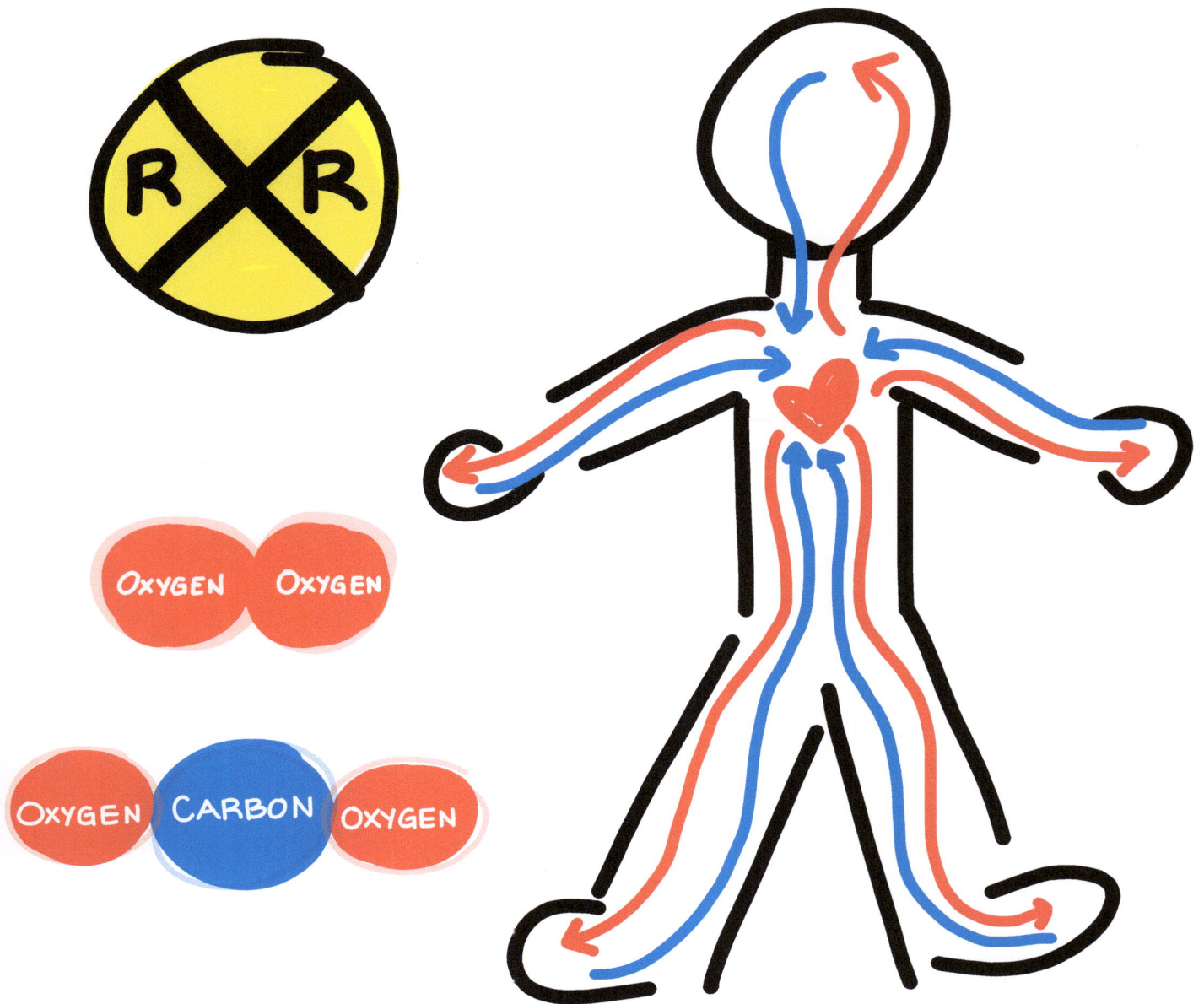

The **circulatory system** is like a railroad, delivering oxygen and chemicals to all cells in the body and taking carbon dioxide out of the body.
The main station is the heart.

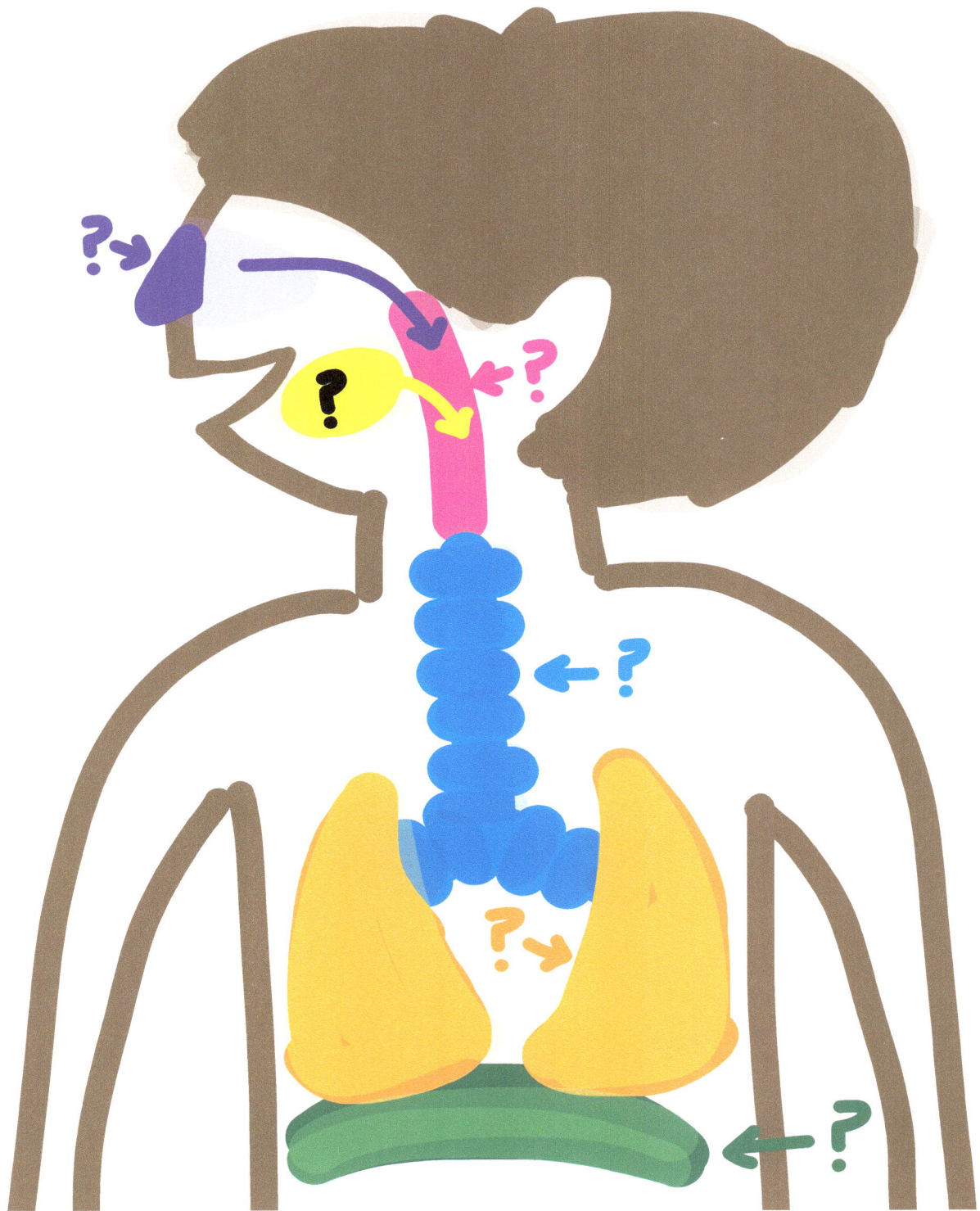

The **respiratory system** works with the **circulatory system**. The **respiratory system** takes in oxygen and releases CO_2.

Digestive System

Sound it Out
1. DI
2. JES
3. TIV

DO YOU LIKE TO EAT?

The digestive system takes in food, digests (or breaks down) the food, absorbs the nutrients we need to be healthy and eliminates the stuff we do not need.

Digestion begins when you chew your food.

This is called **mechanical** digestion.

Mechanical

Sound it Out
1. MEH
2. KAN
3. EH
4. KUL

You are using something, (your teeth), to physically break down the food.

Your **mouth** releases **enzymes** to help break down food while you are chewing. This enzyme in the mouth is called **amylase**.

Amylase

Sound it Out

1. **AM**
2. **EH**
3. **LAYZ**

This is called **chemical** digestion.

Chemical digestion continues in the stomach.

Enzymes

Sound it Out

1. IN
2. ZIMZ

Chemical digestion uses chemicals (enzymes) to break down food in the mouth and in the stomach.

Chemical

Sound it Out

1. KIM
2. EH
3. KUL

Esophagus: E - SOF - UH - GUS

Stomach: STUH - MUK

Small Intestine: SMALL IN - TES - TIN

Large Intestine: LARGE IN - TES - TIN

Liver: LIV - R

Pancreas: PAN - KREE - US

Abdomen: AB - DO - MIN

Food enters the mouth, goes down the esophagus to the stomach. Then it goes to the small intestine and finally to the large intestine, then exits the body. Other organs in your abdomen (or stomach area) are the liver and the pancreas.

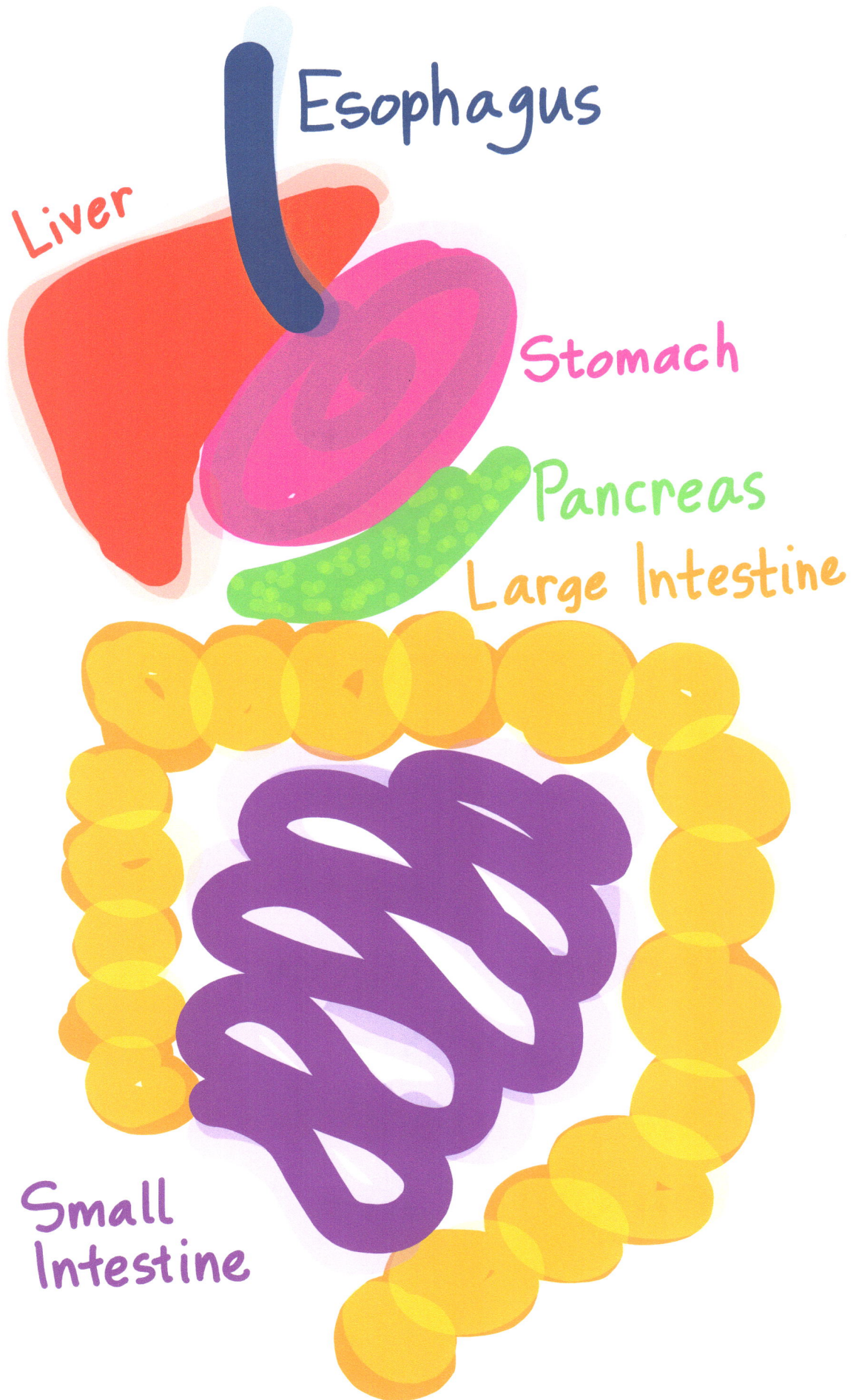

Esophagus

Liver

Stomach

Pancreas

Large Intestine

Small Intestine

The digestive system starts in the mouth with chewing (mechanical digestion).

Chemical digestion occurs when enzymes break down food. This happens in the mouth with amylase and also in the stomach with other enzymes.

Digesting food follows this path: mouth, esophagus, stomach, small intestine, large intestine, and then out of the body.

Integumentary System

The **integumentary system** is your skin, hair and nails. It protects your organs from damage and disease. It also helps regulate (or keep normal) your body temperature.

The **integumentary system** regulates body temperature by sweating to cool your body and shivering to warm your body.

NORMAL

The **integumentary system** is made of your skin, nails and hair.

Skin is a type of **epithelial**.

Epithelial

Sound it Out
1. EP
2. EH
3. THEE
4. LEE
5. UL

Epithelials cover your outer body surfaces and also line internal organs, protecting them from harm.

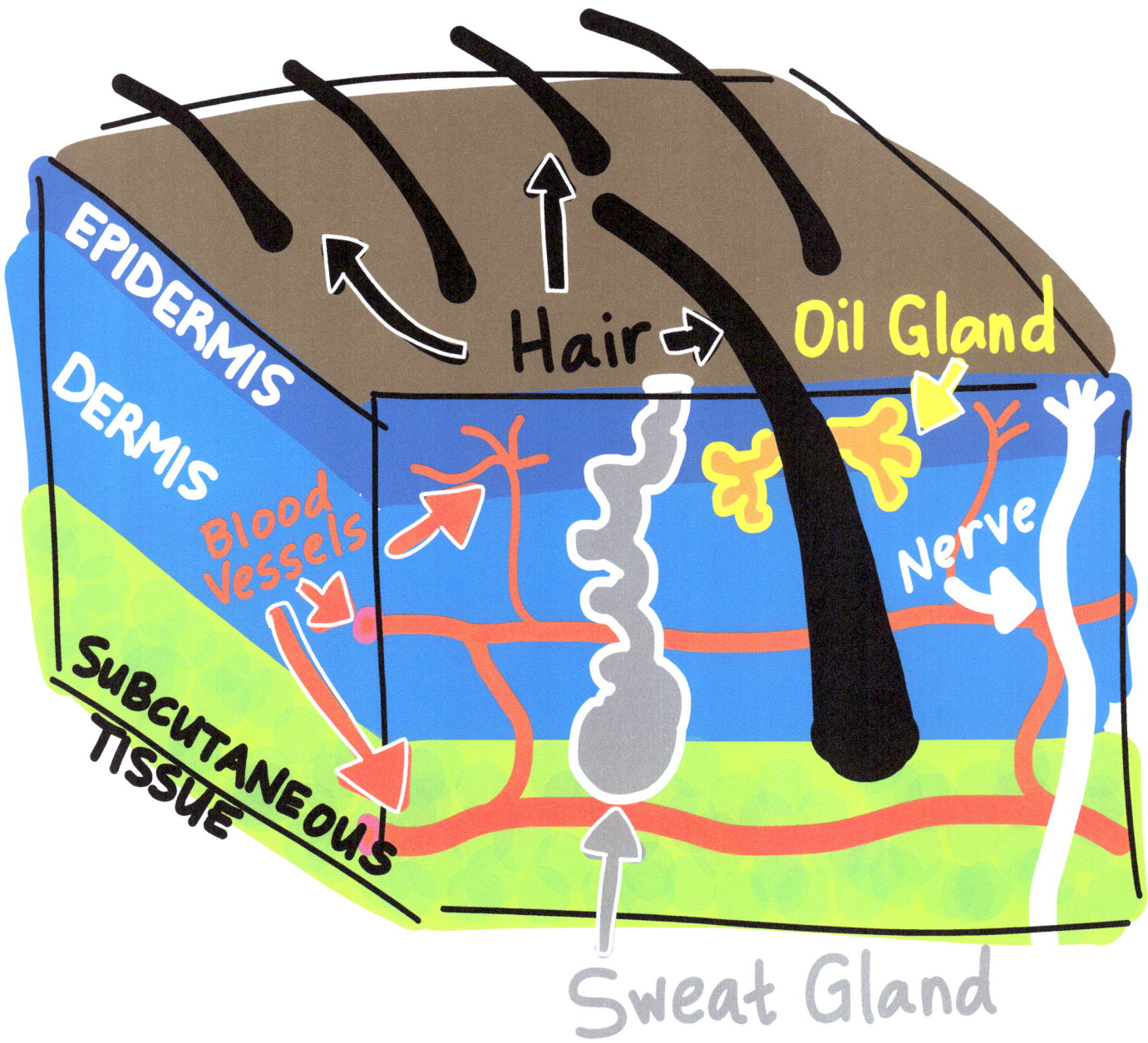

Immune System

Your immune system **protects your body from invaders!**

*Have you ever had a cough or a runny nose***?**

That is your immune system **working to fight off invaders.**

The immune system works with the integumentary and circulatory systems.

The skin (integumentary system) is like a shield and the circulatory system is like a sword.

The **immune system** releases fighters into your blood stream (circulatory system) to attack invaders that get past the shield.

Circulatory System

Blood cells

The immune system works together with the integumentary system and circulatory system to protect your body from invaders.

Skin, hair and nails are all part of the integumentary system.

Skin is a type of epithelial.

The integumentary system regulates body temperature.

The immune system keeps you healthy by fighting off invaders.

If a foreign object gets past the skin or into the blood stream (circulatory system), the immune system sends out fighters to destroy the foreign invaders!

QUESTIONS:

What is the circulatory system?

What is the main station of the circulatory system?

The circulatory system brings _____ to all cells in the body and takes _____ out of the body.

What is the respiratory system?

What are the six parts of the respiratory system?

What is the digestive system?

What is the difference between mechanical and chemical digestion?

What is amylase?

Name the parts of the digestive system.

What is the integumentary system?

What is the immune system?

What happens when invaders get into the circulatory system?

VERY GOOD!!!

Circulatory system

Respiratory system

Digestive system

Mechanical digestion

Chemical digestion

Integumentary system

Immune system

Nose

Mouth

Pharynx

Trachea

Lungs

Diaphragm

Enzymes

Esophagus

Stomach

Small Intestine

Large Intestine

Liver

Pancreas

Abdomen

Epithelial

Cardiology

FOR KIDS

...and Adults too!

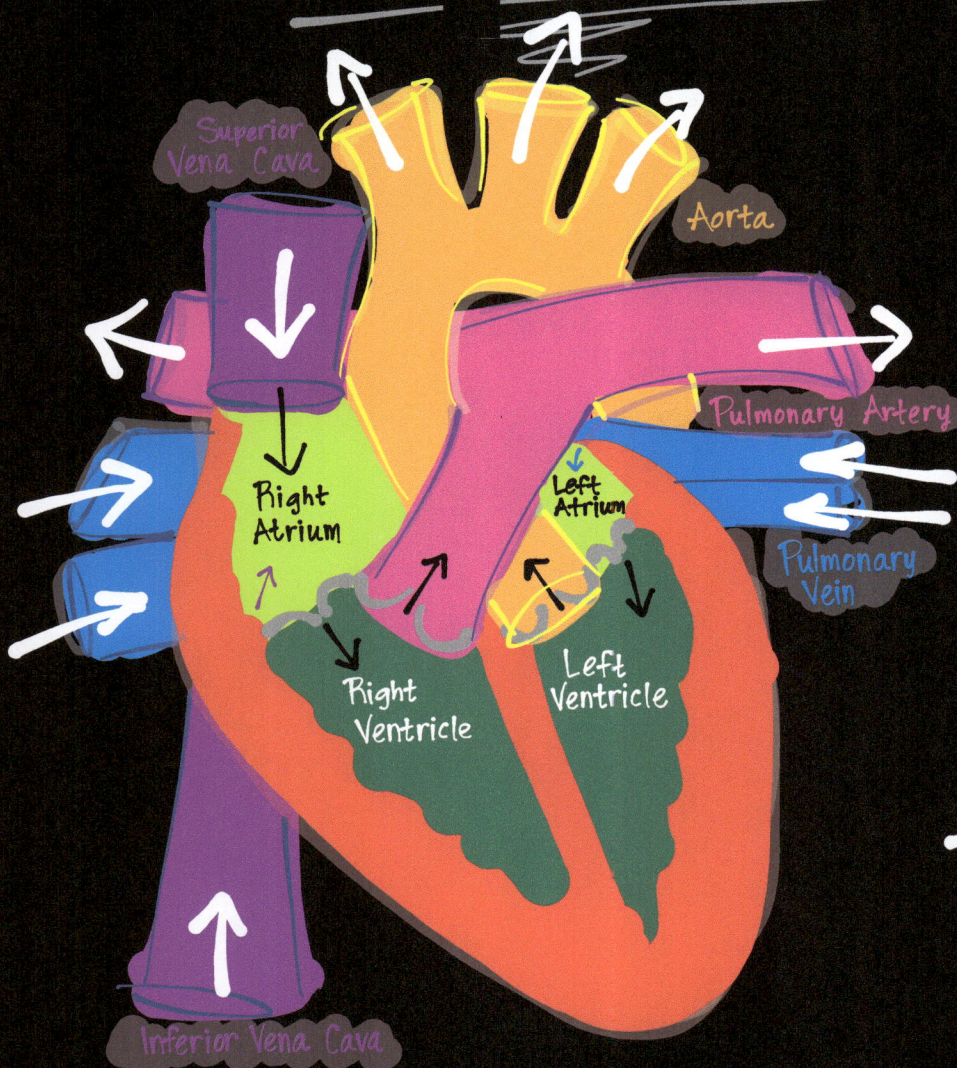

Superior Vena Cava

Aorta

Right Atrium

Left Atrium

Pulmonary Artery

Pulmonary Vein

Right Ventricle

Left Ventricle

Inferior Vena Cava

By: April Chloe Terrazas

Book 7 of the Super Smart Science Series
Ages 0 - 100

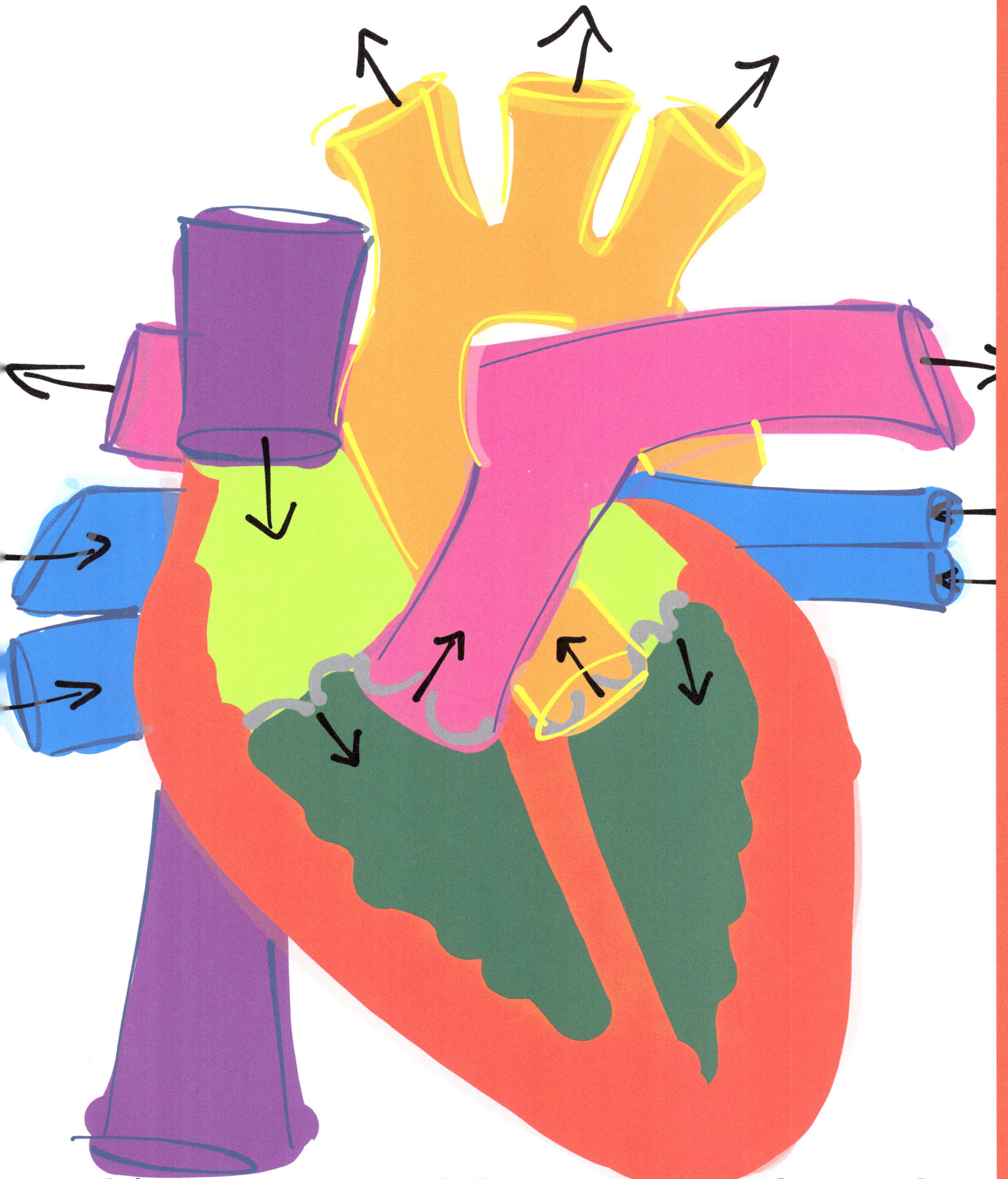

This SUPER AWESOME Book Belongs to:

_____

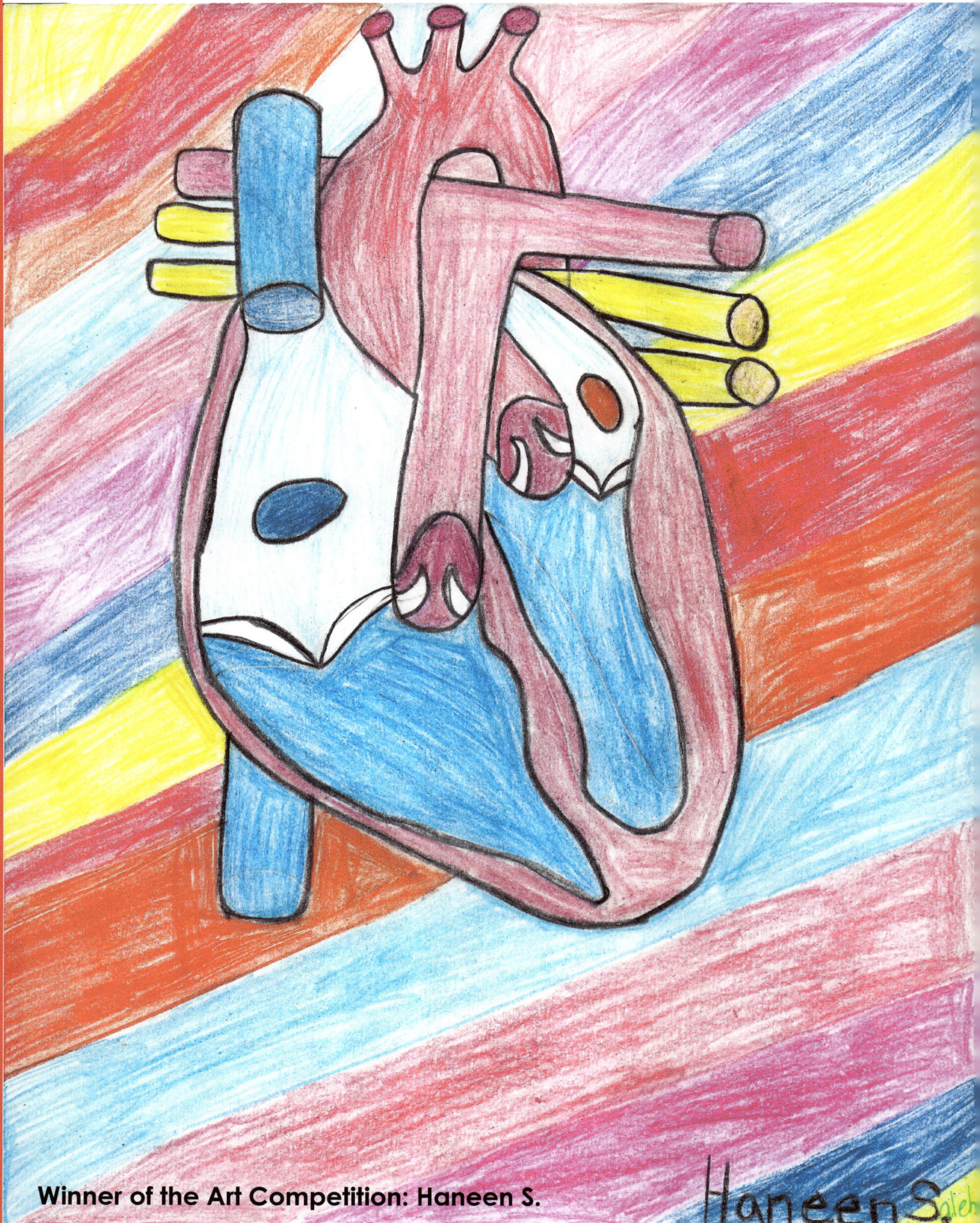

Winner of the Art Competition: Haneen S.

Cardiology FOR KIDS

...and Adults too!

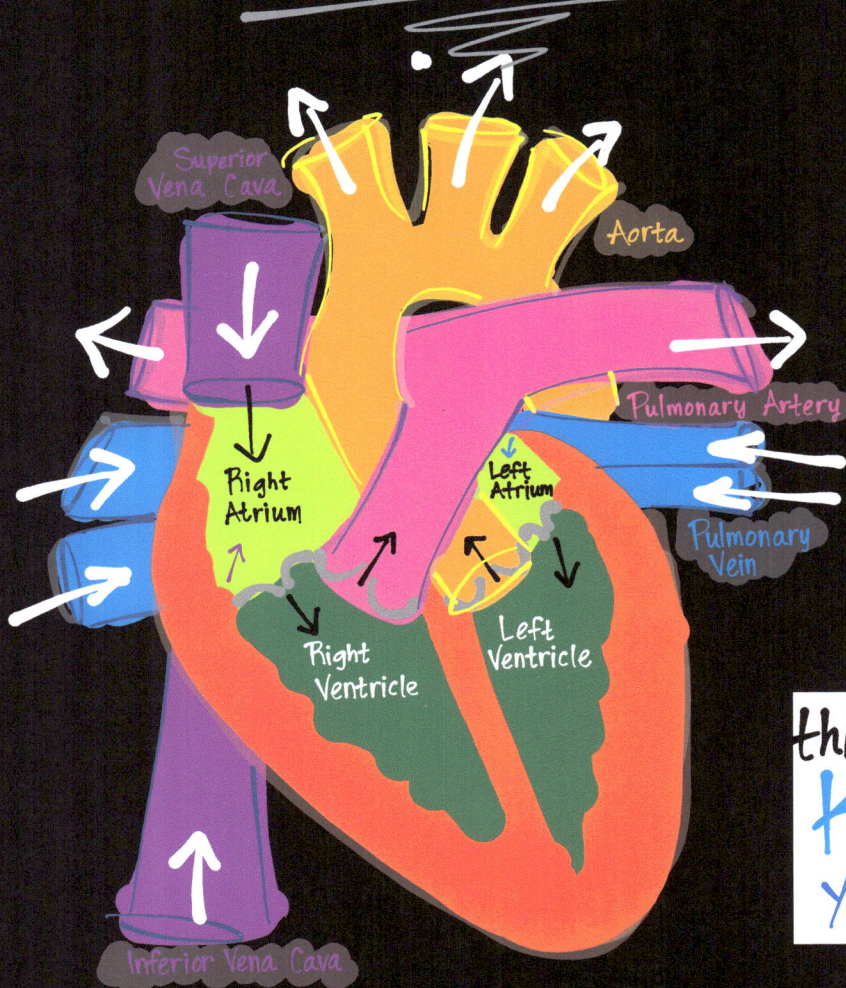

Superior Vena Cava

Aorta

Pulmonary Artery

Right Atrium

Left Atrium

Pulmonary Vein

Right Ventricle

Left Ventricle

Inferior Vena Cava

By: April Chloe Terrazas

this book is dedicated to **Kenny Kent!** You are AWESOME.

Cardiology FOR KIDS ...and Adults too!   By: April Chloe Terrazas, BS University of Texas at Austin.
Copyright © 2014 Crazy Brainz, LLC

Visit us on the web! **www.Crazy-Brainz.com**

Cover design, illustrations and text by: April Chloe Terrazas

Cross your right hand over your chest. That is where your heart is located.

Make a fist with your hand. That is about the size of your heart.

Take two fingers and touch the side of your neck.

Can you feel your heartbeat?

When you finish this book, you will know what every colored part of the heart does!

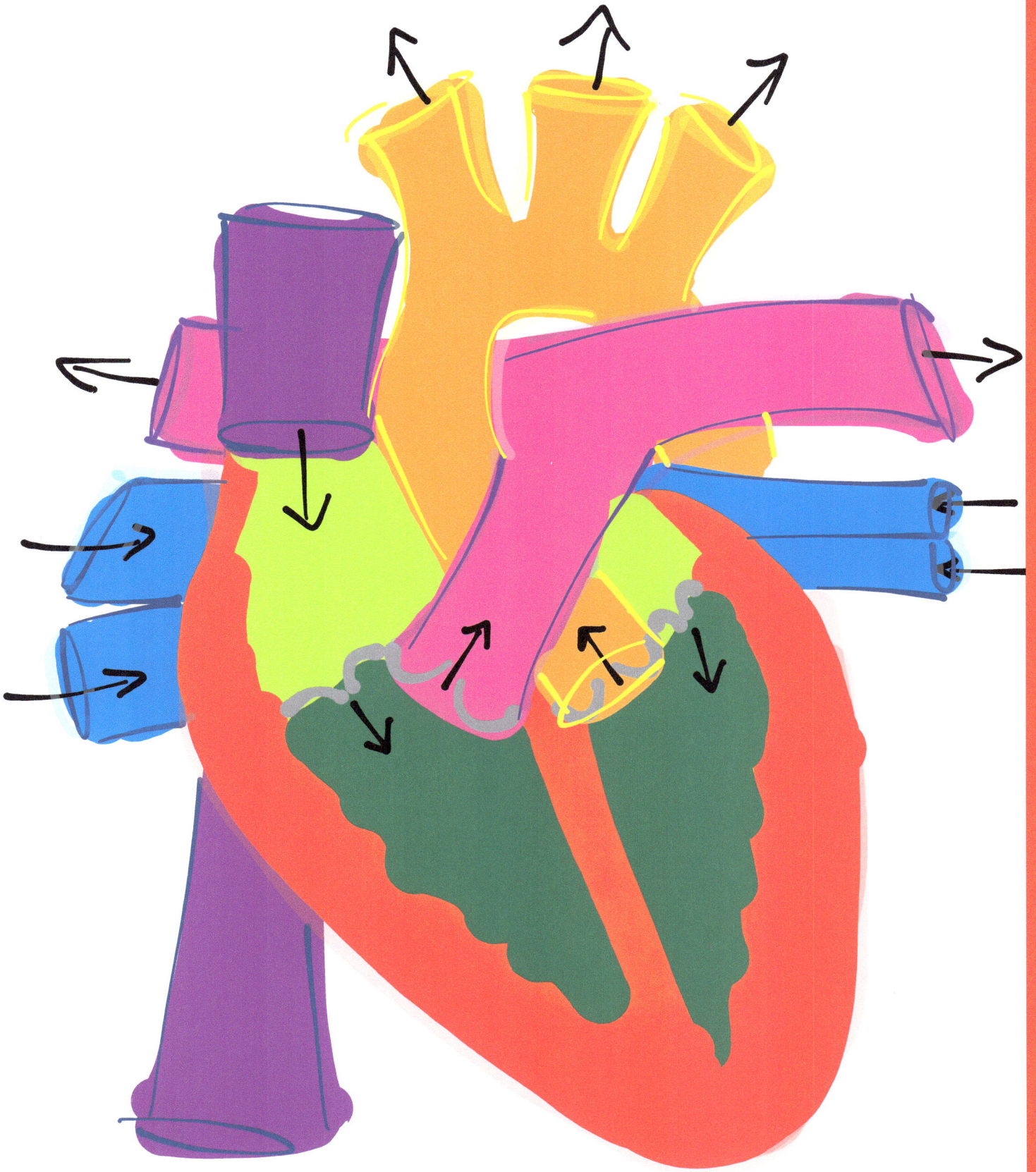

The main station in the Circulatory System Railroad is the heart.

The "tracks" on the railroad are:

Arteries: AR-TER-EEZ

Arterioles: AR-TER-EE-OLZ

Arteries are large "tracks" (blood vessels) that leave directly from the main station, the heart. Arterioles are smaller "tracks" that lead from the larger arteries to an even smaller blood vessel.

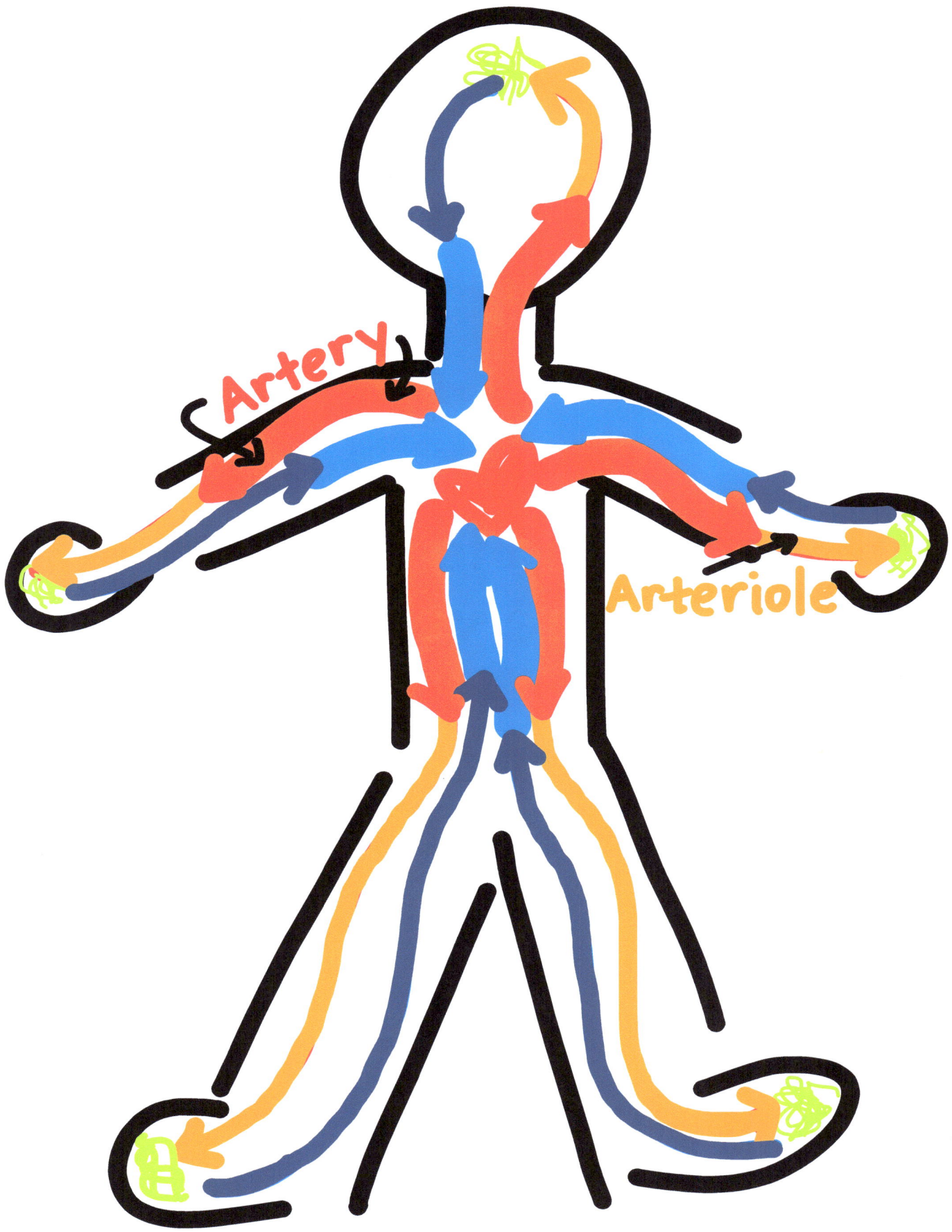

Artery

Arteriole

Capillaries: CAP-IL-AIR-EEZ

Venules: VIN-ULZ

Veins: VAYNZ

Arterioles lead to capillaries. Capillaries are very small!

Venules connect on the other side to the tiny capillaries.

Veins are larger "tracks," like arteries. Veins bring blood back to the heart to get more oxygen.

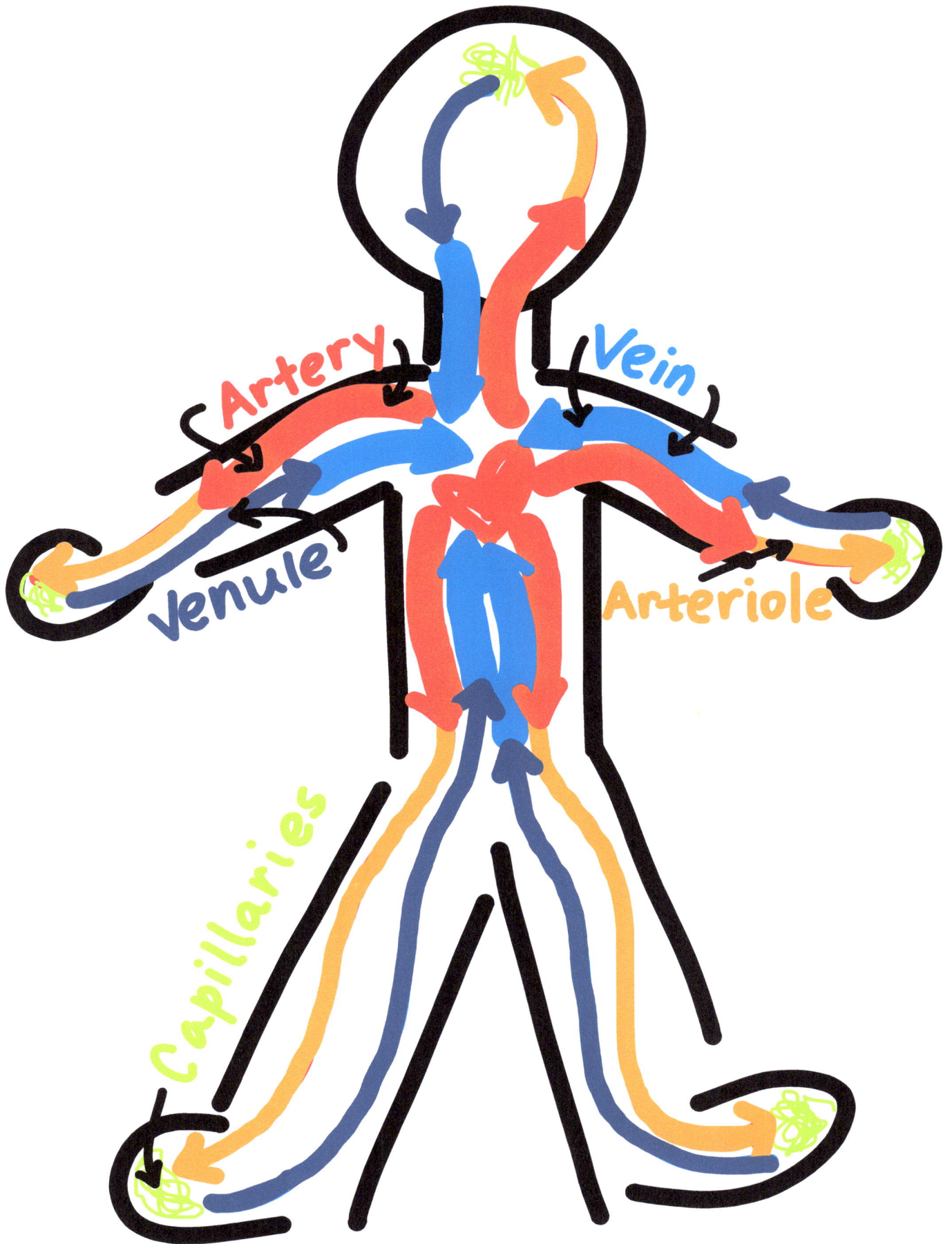

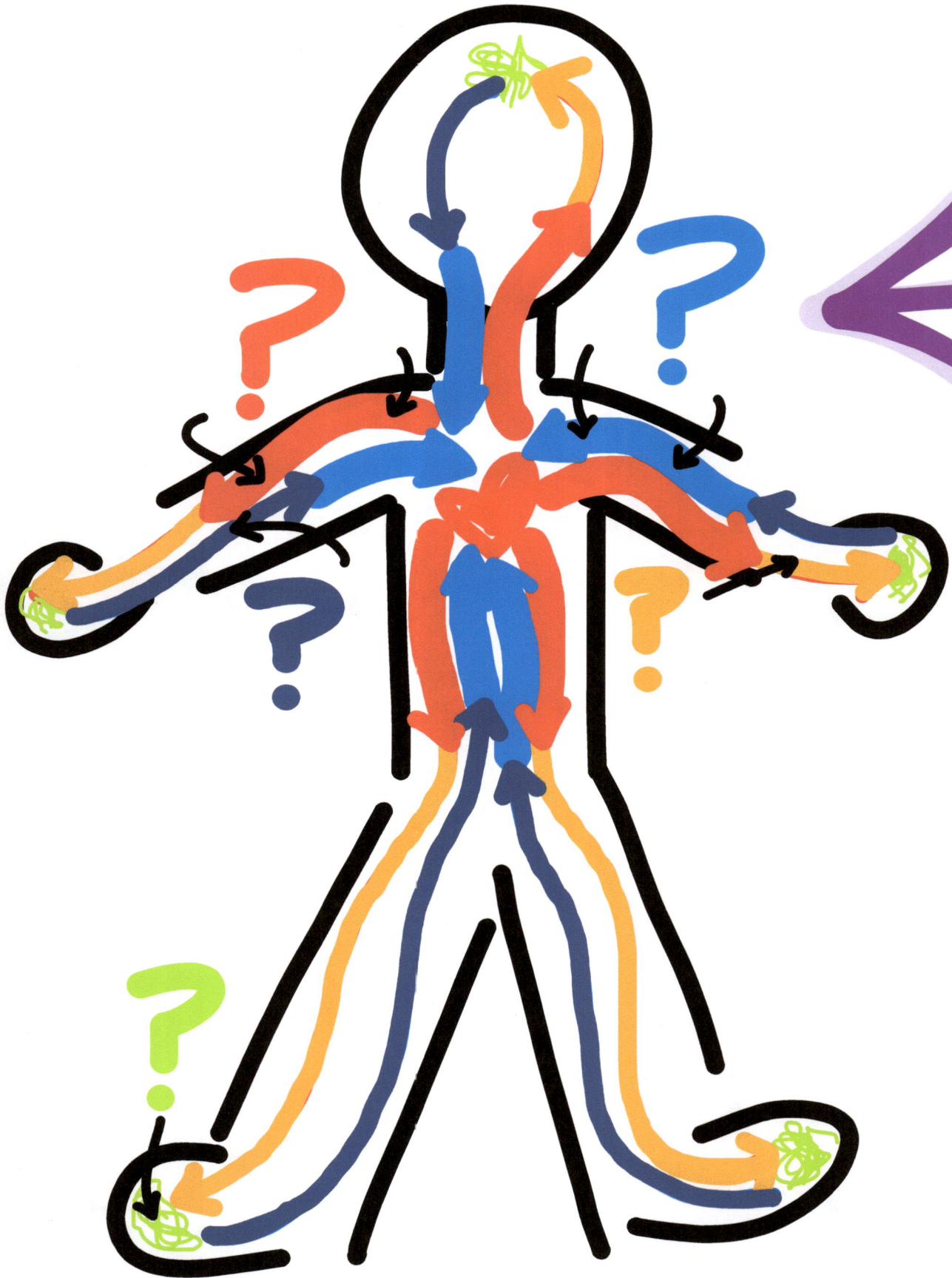

What is the main station of the circulatory system railroad?

What are the names of the "tracks" on the circulatory system railroad?

Where is your heart located?

What is the size of your heart?

Where can you feel your heartbeat?

The right side of the heart receives blood from the body and sends it to the lungs.

When this blood reaches the lungs, it exchanges CO_2 (carbon dioxide) for O_2 (oxygen).

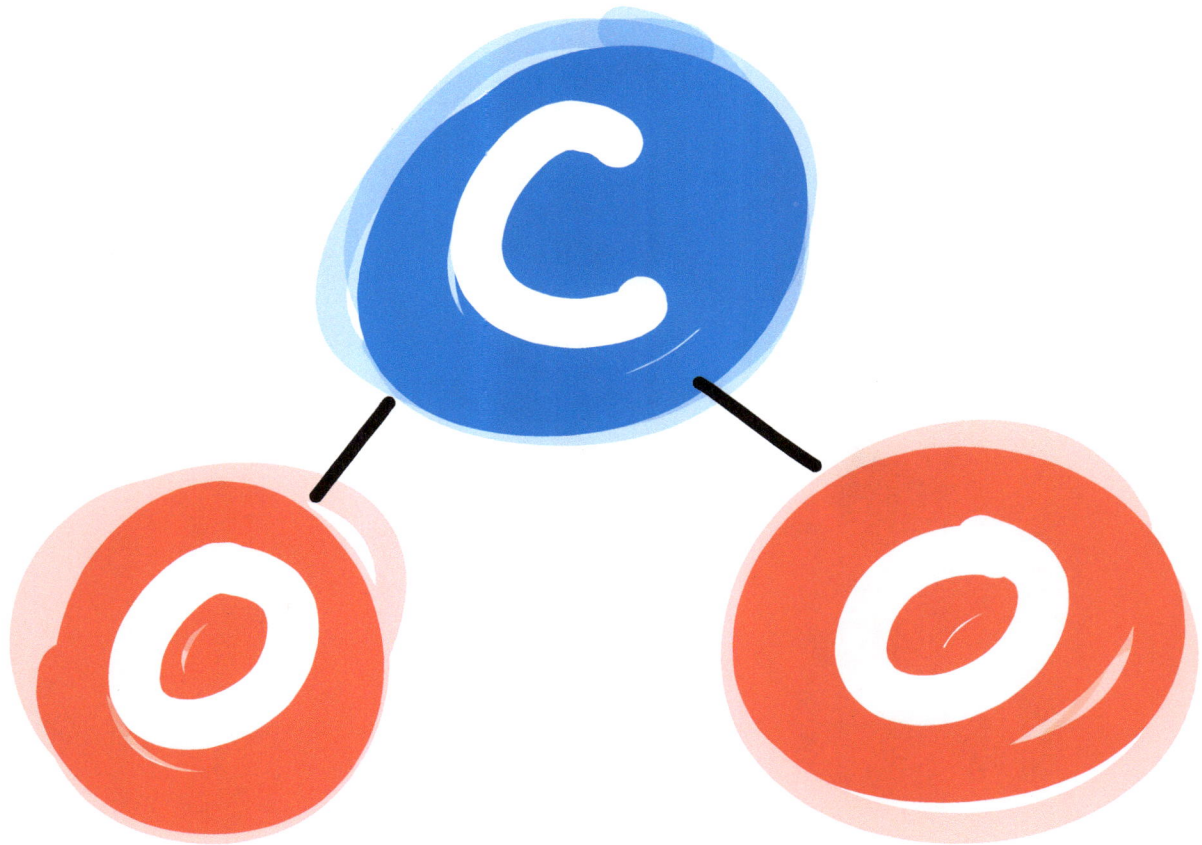

What is the name of this molecule?

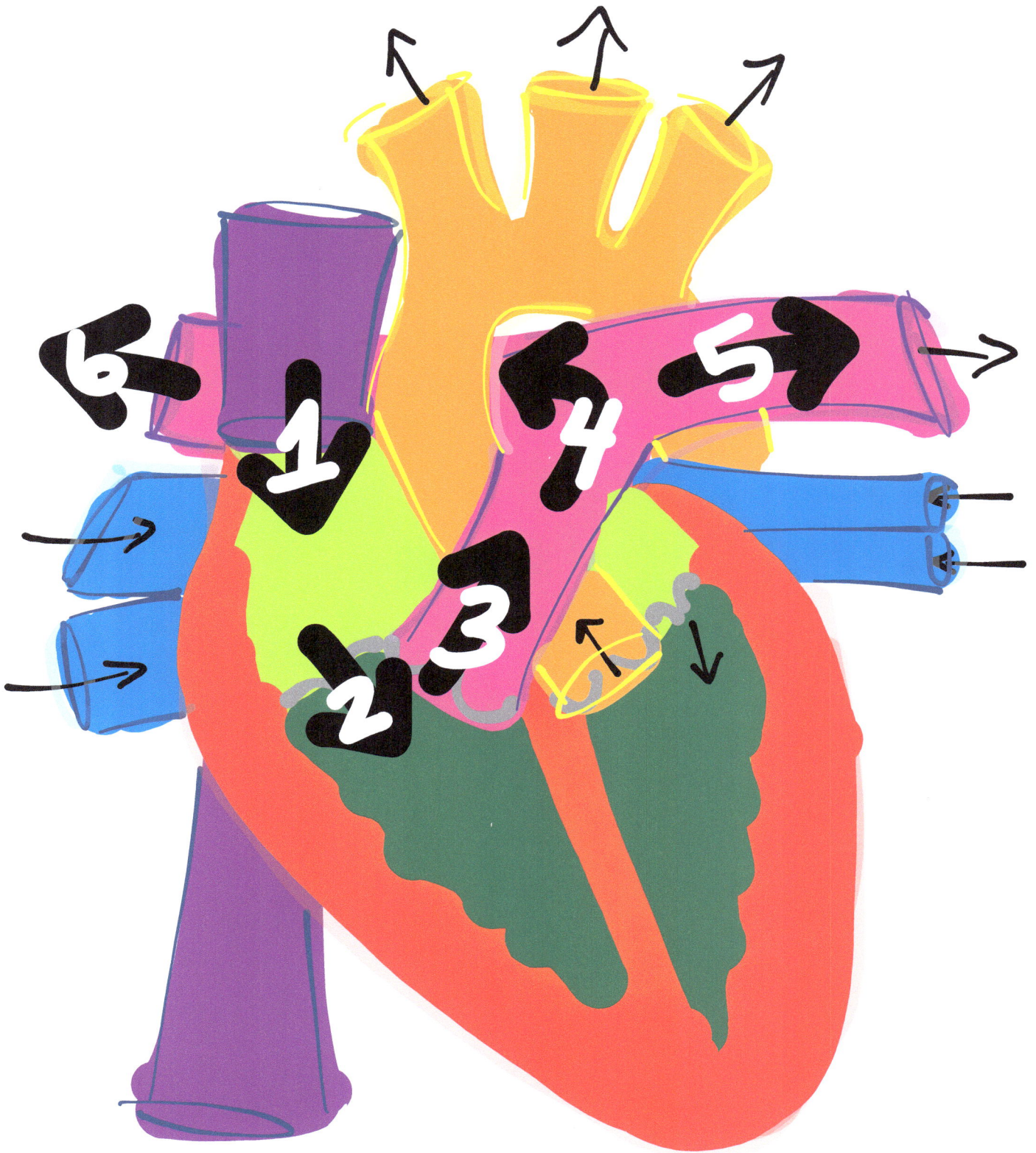

#5 and #6 are heading toward the lungs.

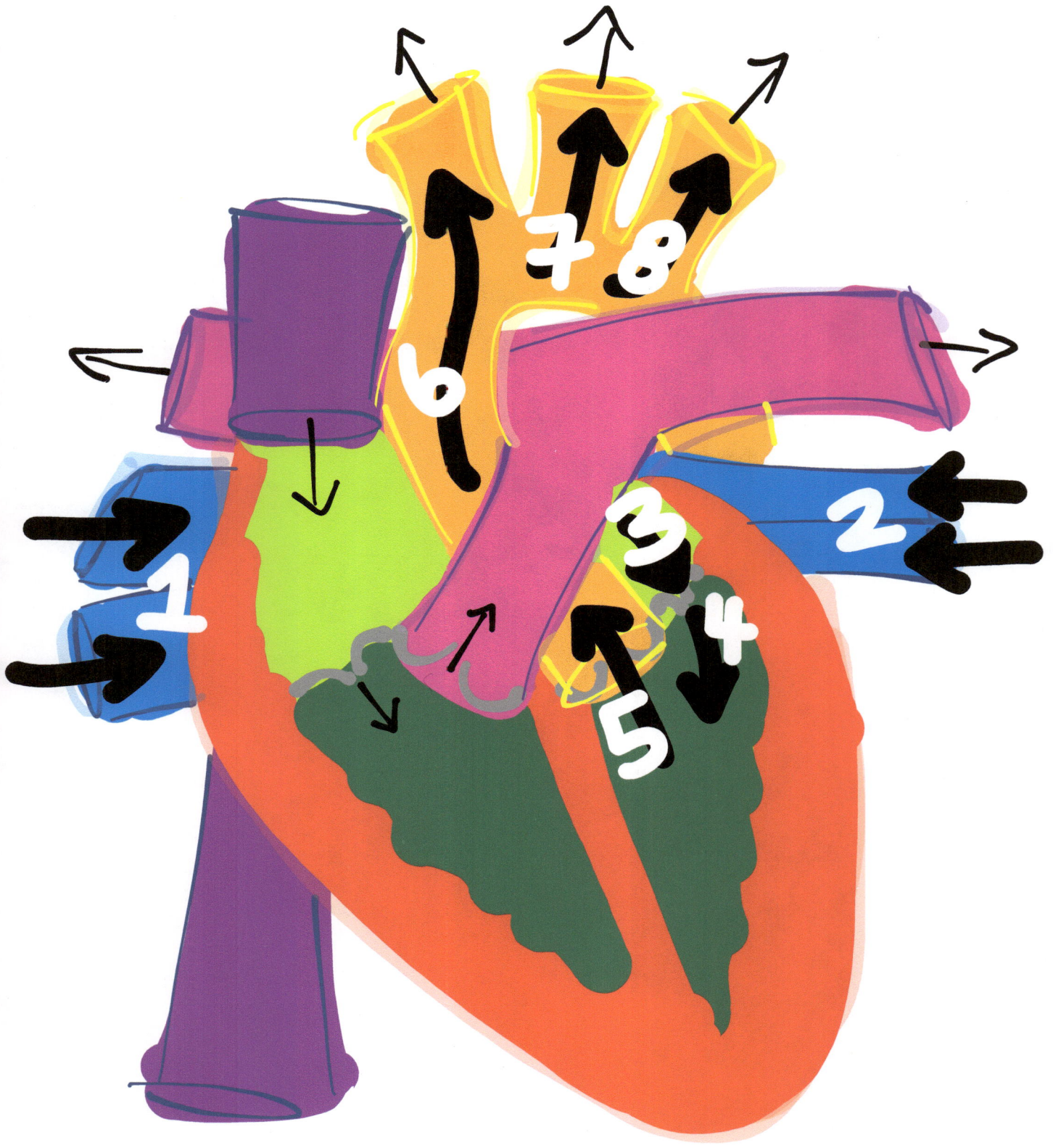

The blood returning to the heart from the lungs (through #1 and #2) now has O_2 (oxygen).

The newly oxygenated blood coming from the lungs enters the left side of the heart (#3, #4 and #5).

It is then pumped back out, delivering O_2 (oxygen) to all of the cells in the body.

#6, #7 and #8 are heading out all over the body.

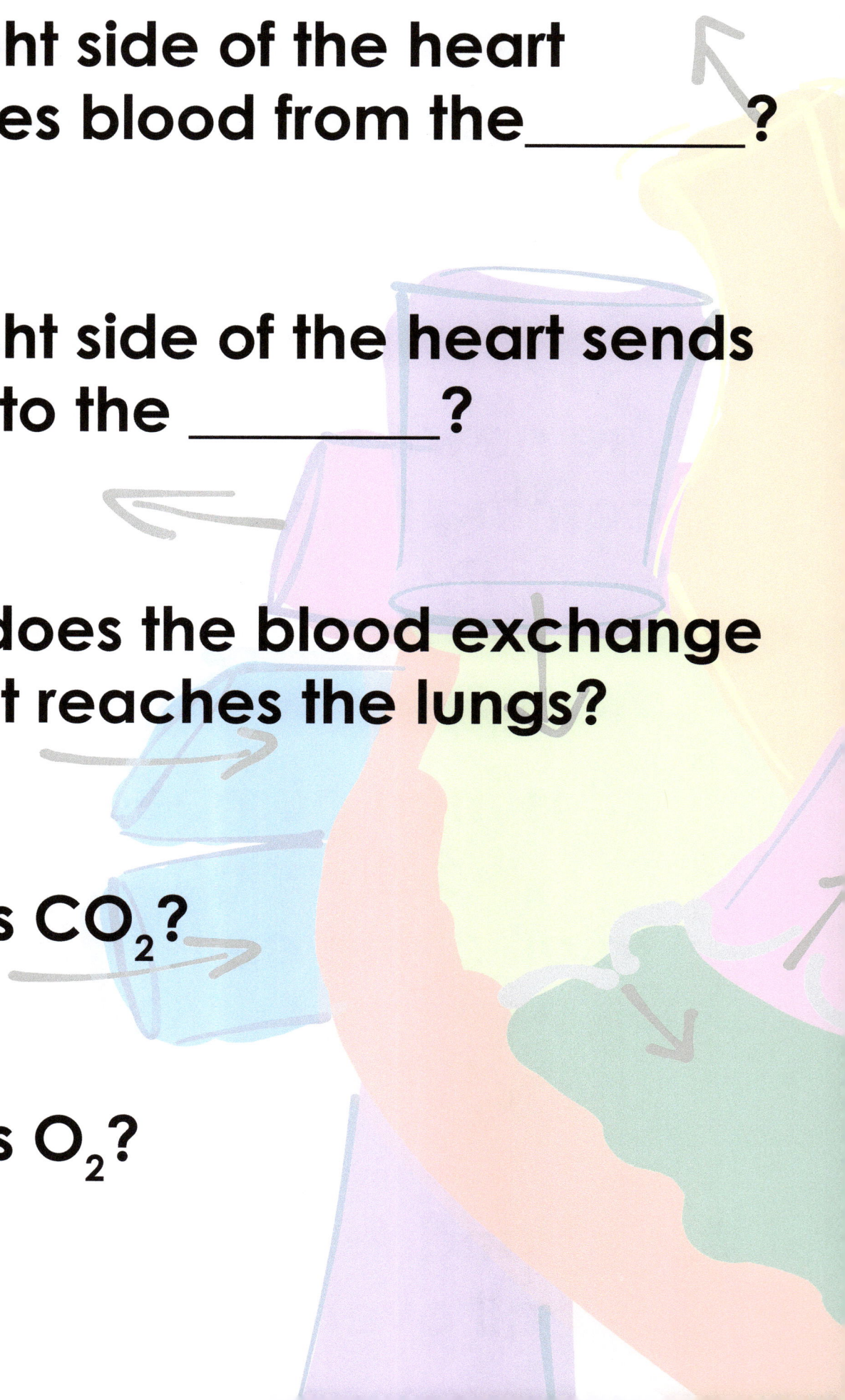

The right side of the heart receives blood from the _____?

The right side of the heart sends blood to the _____?

What does the blood exchange when it reaches the lungs?

What is CO_2?

What is O_2?

The left side of the heart receives blood from the _____?

This blood is now carrying _____ that it received from the lungs.

The blood leaves the left side of the heart and travels all over the _____.

Well done!

The heart has four main areas called chambers.

The 2 chambers on top are called atria (an individual chamber is called an atrium).

There is a left and a right atrium.

Atria

Sound it Out

1. A
2. TREE
3. UH

Atrium

Sound it Out

1. A
2. TREE
3. UM

The atria receive blood from the body and from the lungs.
Blood always enters the atria first.

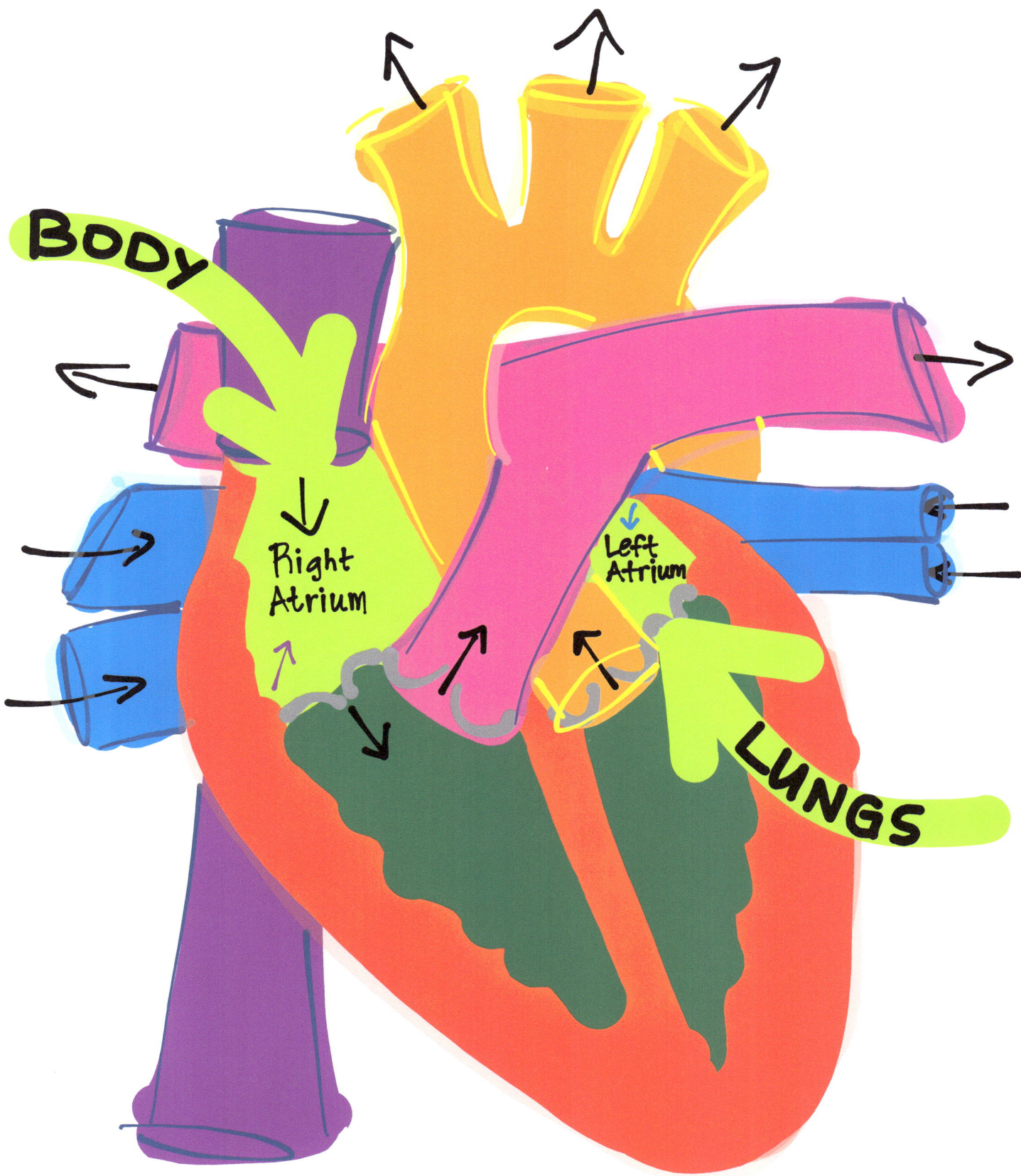

BODY

Right
Atrium

Left
Atrium

LUNGS

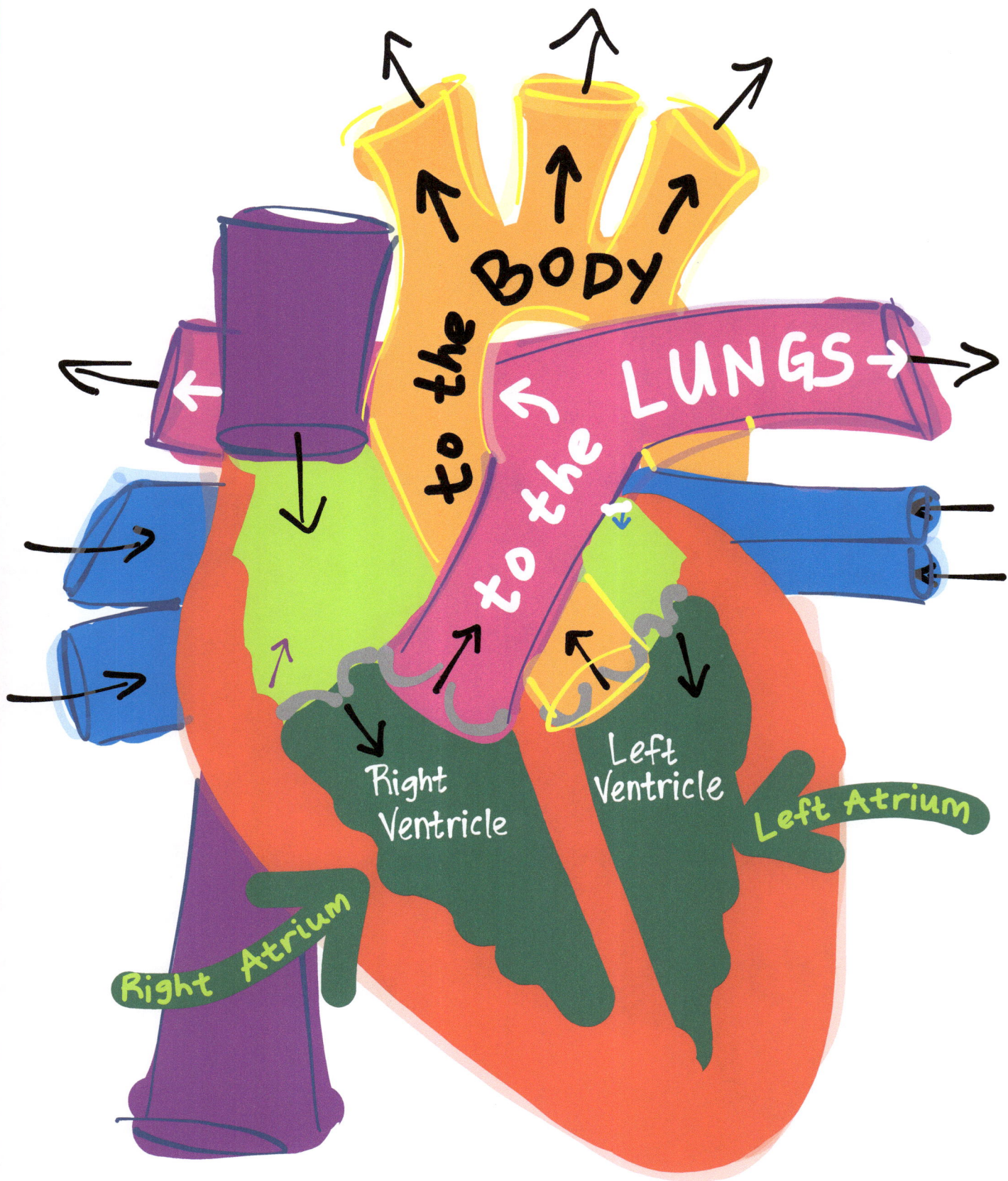

to the BODY

to the LUNGS →

to the L...

Right Ventricle

Left Ventricle

Left Atrium

Right Atrium

The 2 chambers on the bottom are called **ventricles**.

Ventricles

Ventricles receive blood from the **atria** and then move blood out of the heart, to the body and lungs.

You have a left **ventricle** and a right **ventricle**.

The right atrium receives oxygen-poor blood from the body through the superior and inferior vena cava and passes it along to the right ventricle. The vena cava is a large vein.

Superior (SOO-PEER-EE-OR) = top.
Inferior (IN-FEER-EE-OR) = bottom.

The right ventricle then pushes the blood out of the heart and toward the lungs where it picks up oxygen.

Vena Cava

Sound it Out

1. VEE
2. NUH
3. CAV
4. UH

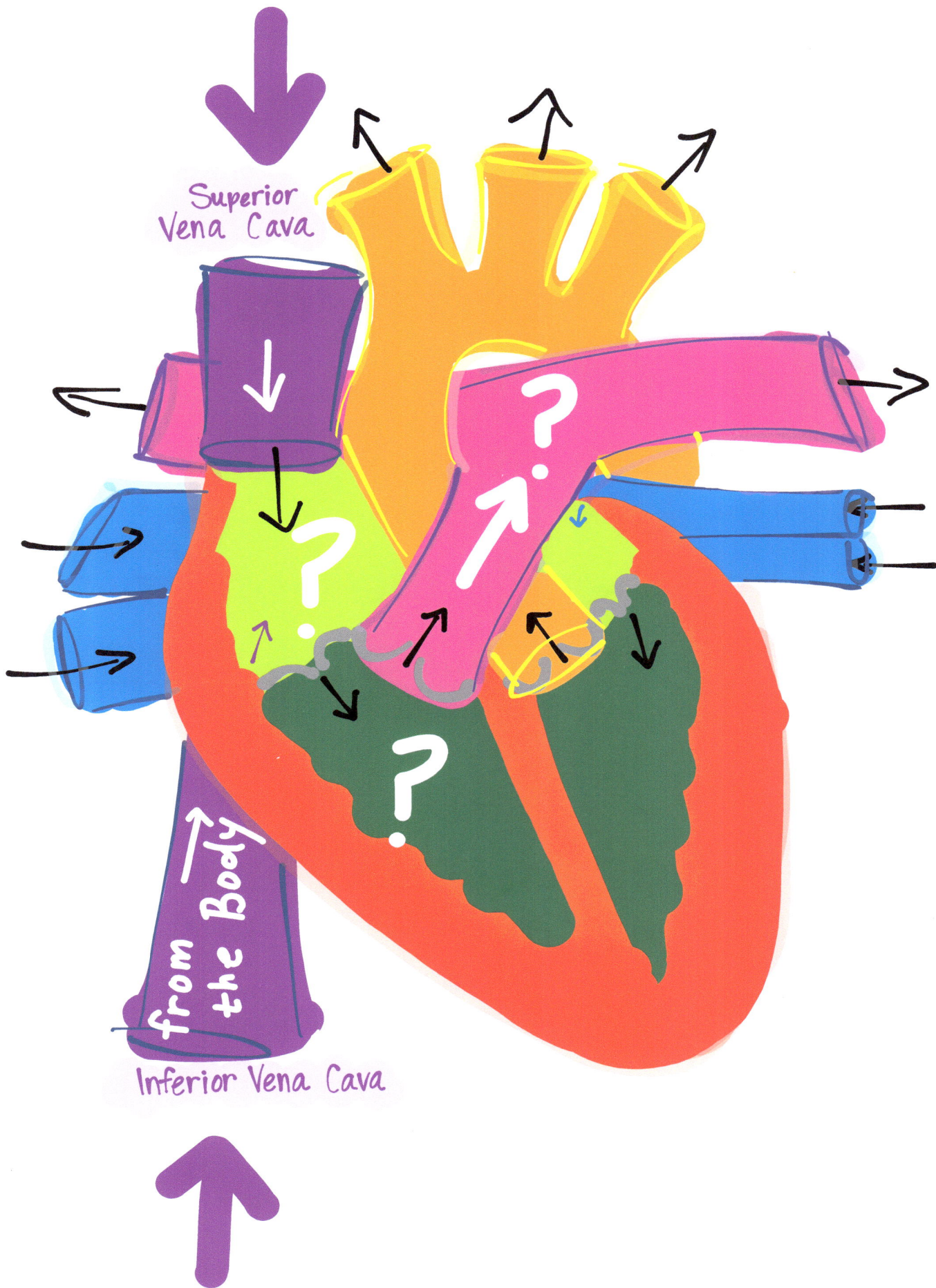

Superior
Vena Cava

from
the Body

Inferior Vena Cava

How many chambers are in the heart?

The chambers in the top of the heart are called _____.

The chambers in the bottom of the heart are called _____.

Which chamber does blood enter first?

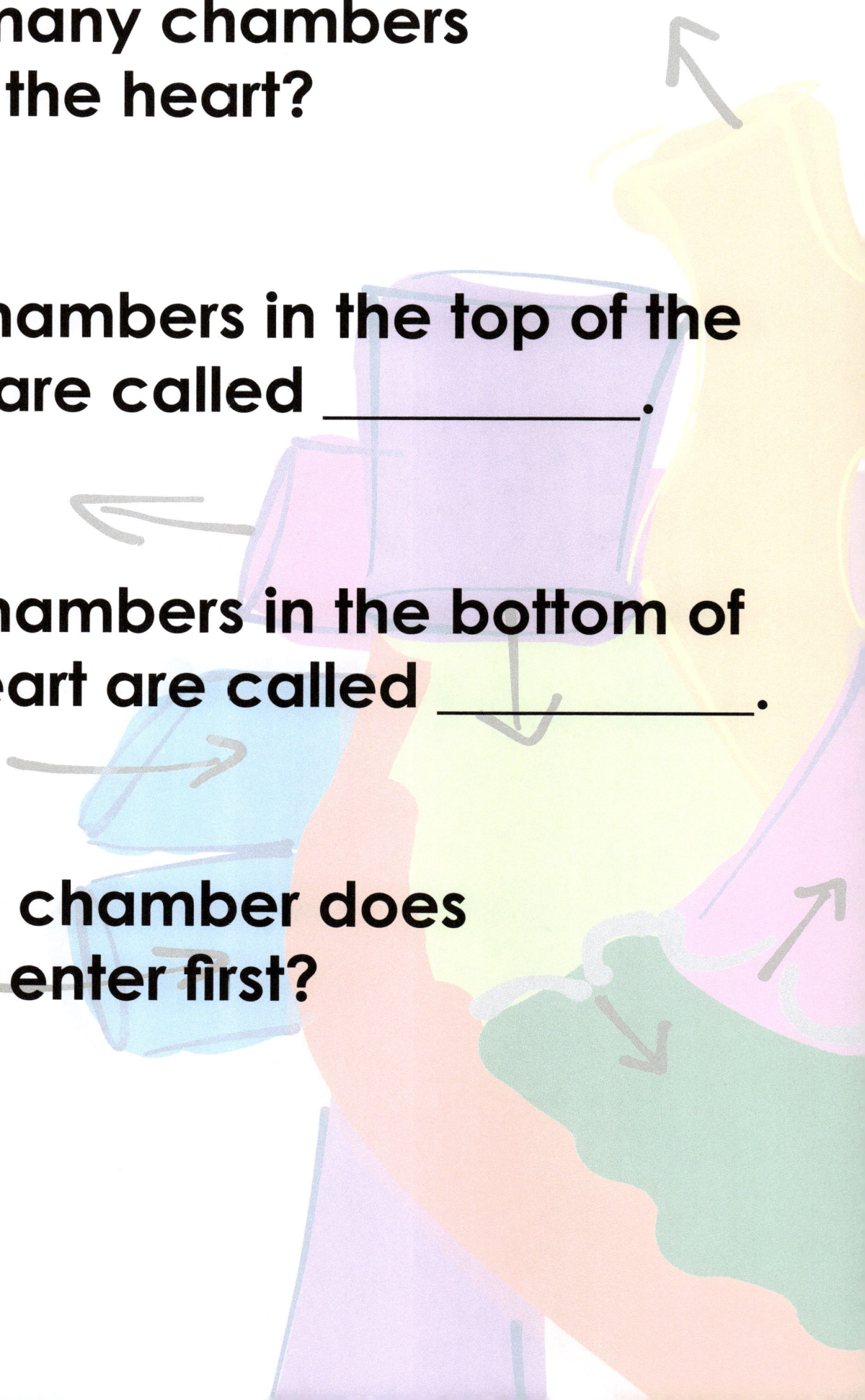

What does SUPERIOR mean?

What does INFERIOR mean?

What is the name of the large vein that brings blood to the right atrium?

Fantastic!

The blood leaves the right ventricle and travels through the pulmonary artery toward the lungs.

Pulmonary is a word you use when speaking about the lungs.

The pulmonary artery goes to the lungs.

Pulmonary

Sound it Out

1. PUL
2. MUH
3. NER
4. EE

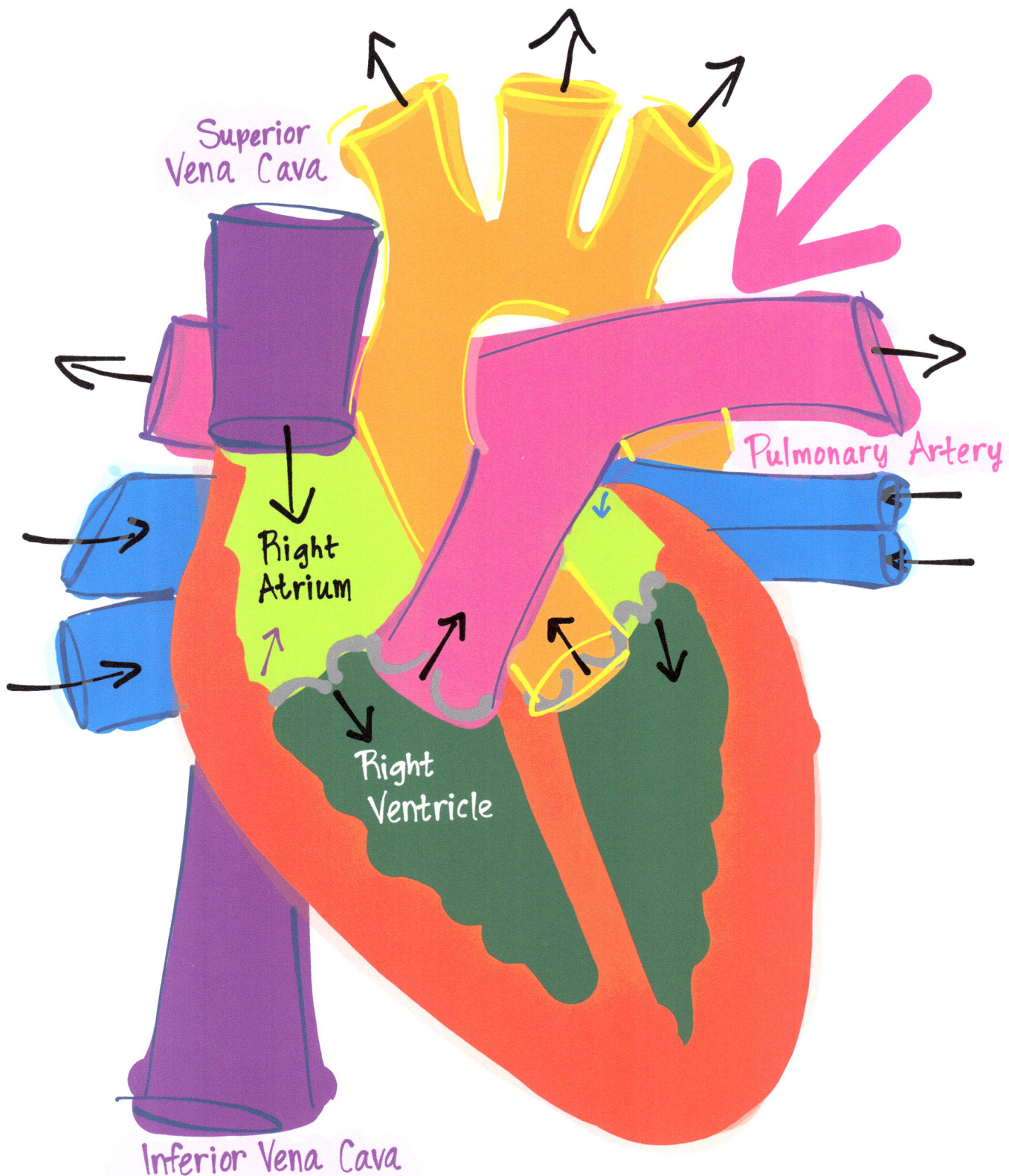

Superior Vena Cava

Pulmonary Artery

Right Atrium

Right Ventricle

Inferior Vena Cava

The blood receives O$_2$ (oxygen) from the lungs and travels back to the heart.

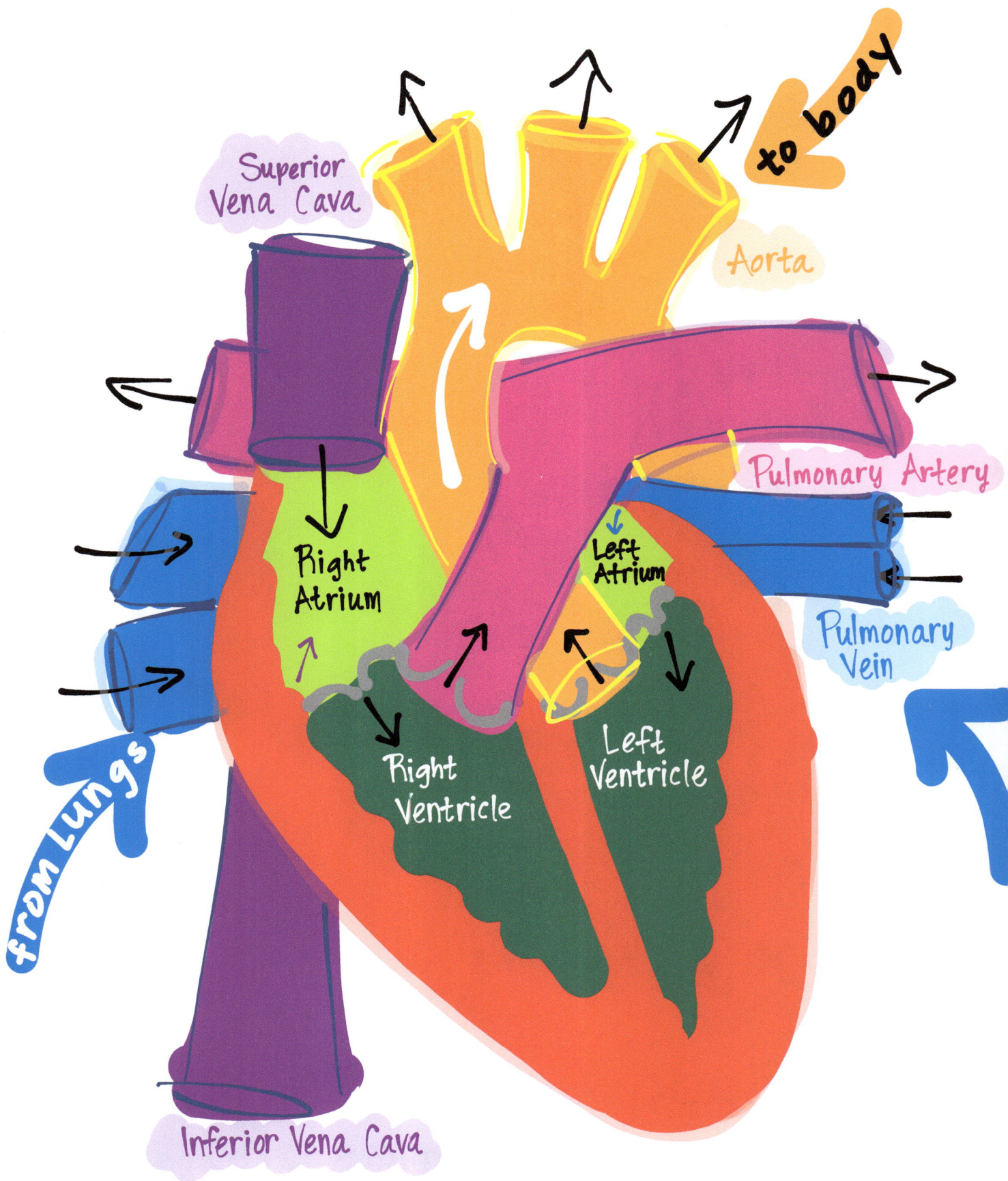

After getting oxygen in the lungs, the blood comes back through the pulmonary vein and into the left atrium of the heart.

The blood moves to the left ventricle where it is pushed out to the aorta and to the body. The aorta is the main artery of the body.

Aorta

Sound it Out

1. A
2. OR
3. TUH

What does the word "pulmonary" refer to?

Blood follows the path from the superior and inferior _____, to the right _____, then to the right _____.

When the blood leaves the right ventricle, what does it travel through? *(hint: it's pink!)*

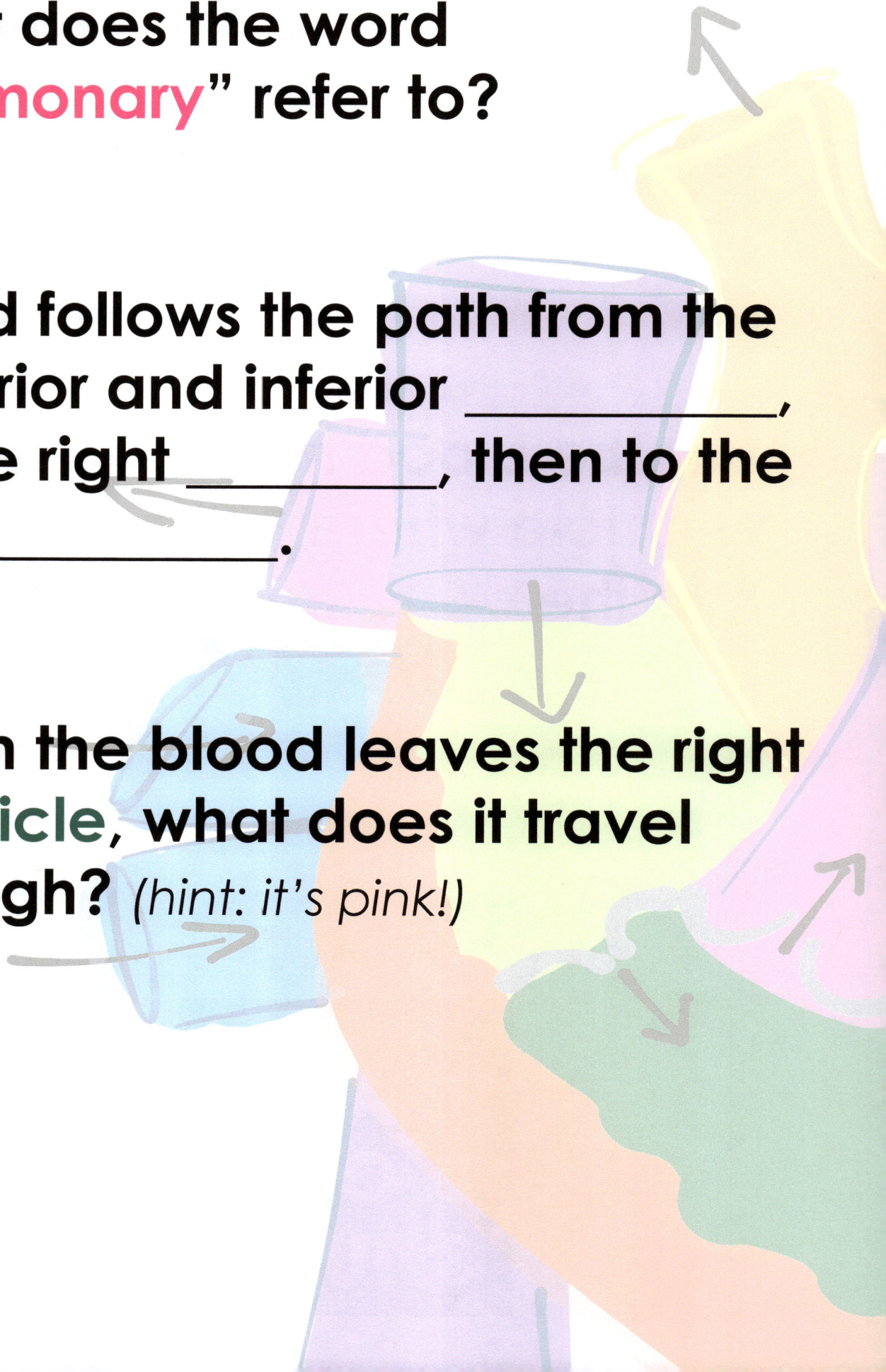

What does blood pick up in the lungs?

After blood gets _____ from the lungs, it travels back to the heart through the _____ _____.

What is the name of the main artery of the body?

(hint: it's orange!)

Very good!

Did you know that your heart has doors just like a house?

The blood stops in the right atrium "room", goes through the door (tricuspid valve) to the right ventricle "room", stops, goes through the door (pulmonary valve) to the pulmonary artery.

The heartbeat sound is made by the closing of the valves.

Tricuspid

Sound it Out

1. TRI
2. KUS
3. PID

Mitral

Sound it Out

1. MI
2. TRUL

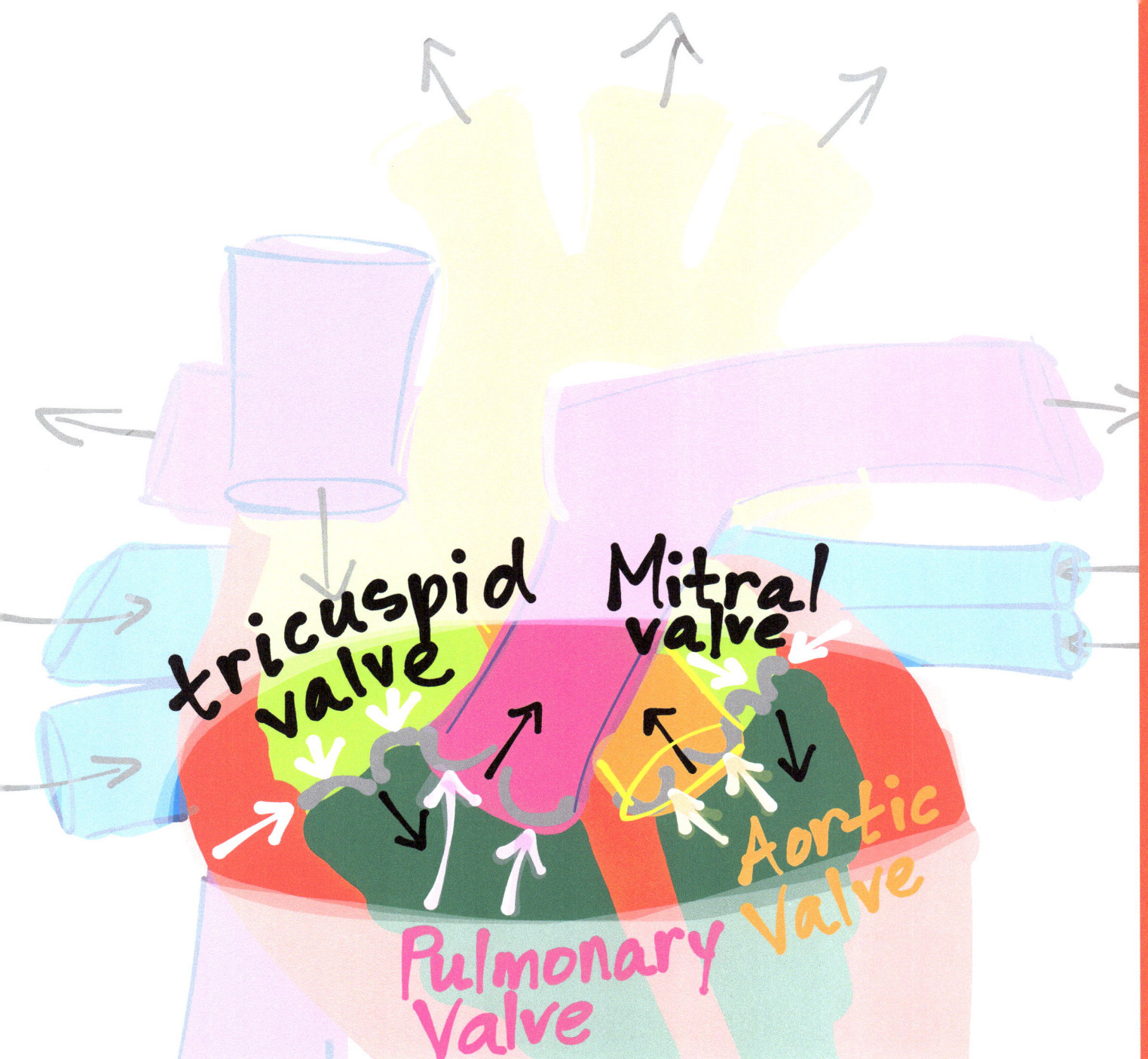

tricuspid valve

Mitral valve

Pulmonary Valve

Aortic Valve

Blood comes back from the lungs into the left atrium "room," stops, goes through the door (mitral valve) to the left ventricle "room," stops, and goes through the door (aortic valve) to the aorta.

Heart contraction starts at the SA node (sinoatrial node). It is located in the right atrium.

This sends a signal (in yellow) to the AV node (atrioventricular node). It is located between the right atrium and right ventricle.

The AV node continues the signal down the purkinje fibers which causes the ventricles to contract, moving blood to the body and lungs!

When the heart relaxes, it fills with blood again.

Purkinje: PUR-KIN-JEE

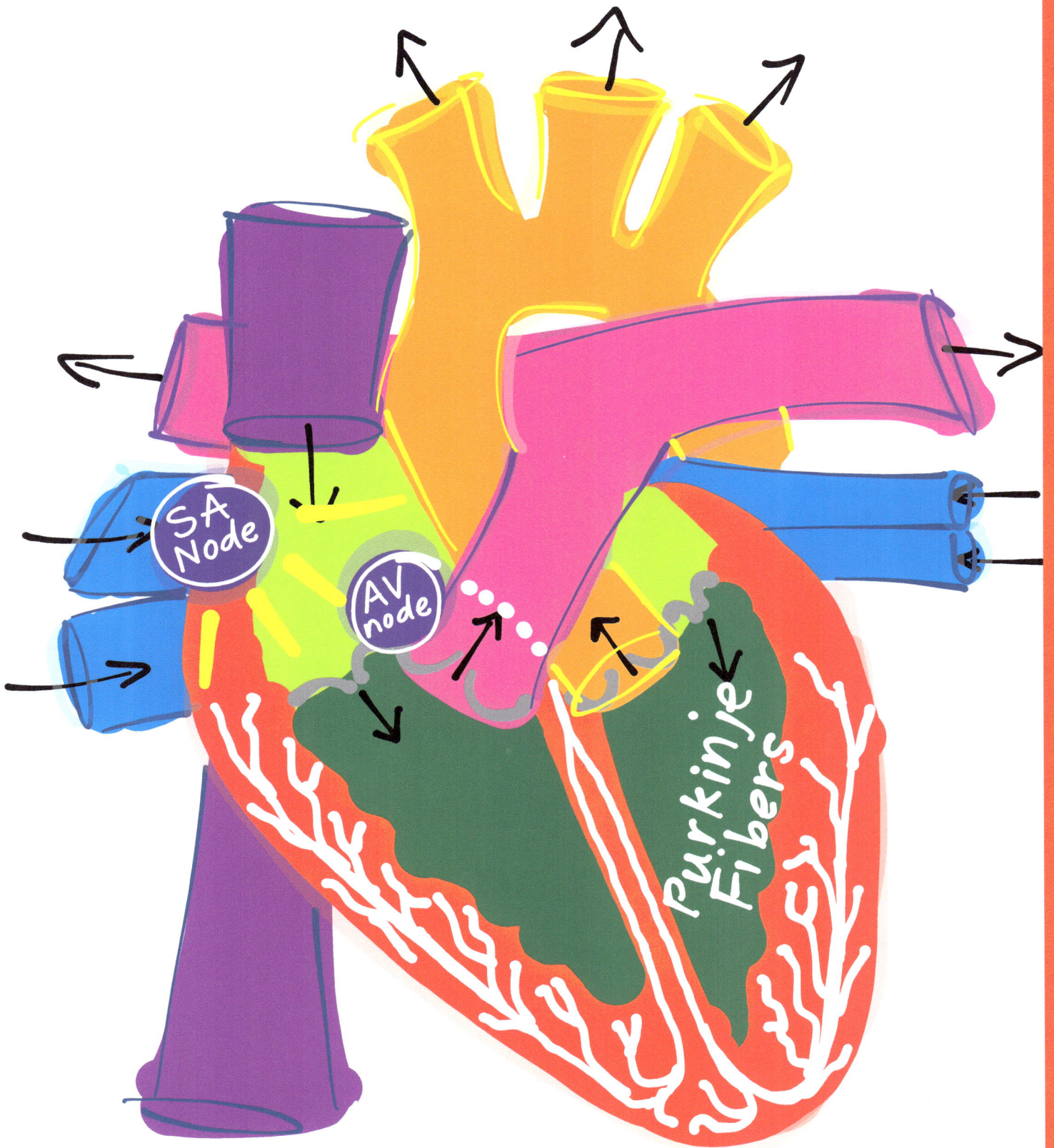

SA Node

AV node

Purkinje Fibers

Sinoatrial: SI-NO-A-TREE-UL

Atrioventricular:
A-TREE-O-VIN-TRIK-U-LER

What are the "rooms" of the heart?

What are the names of the valves that go from the atria to the ventricles? *(hint: they are grey!)*

What are the names of the valves that go from the ventricles leading out of the heart? *(hint: one is pink, one is orange)*

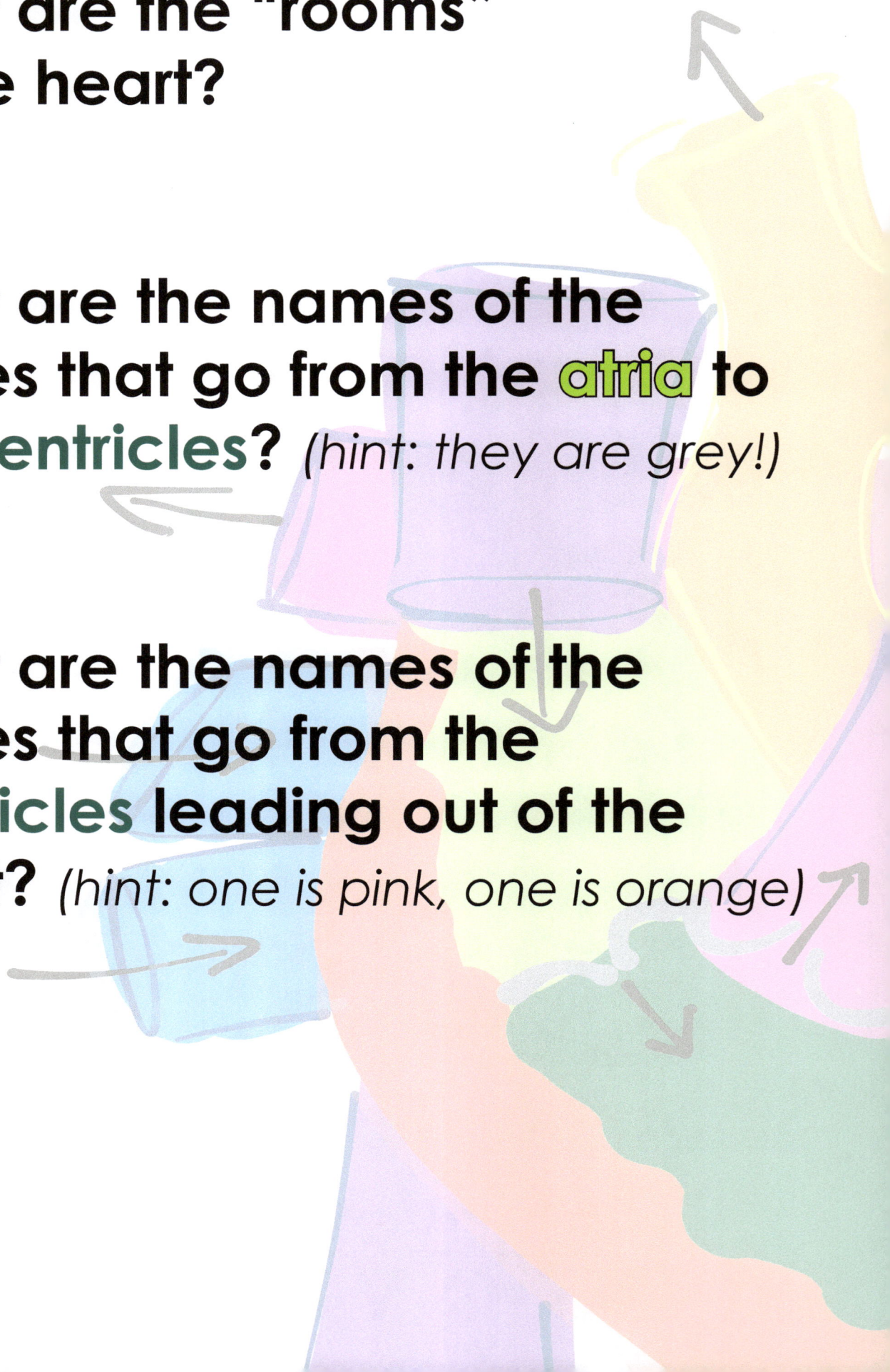

Where does the heartbeat start?

The signal travels to the _____ _____, located between the atrium and ventricle.

What is the name of the fibers that spread around the heart, causing it to contract?

What makes the heartbeat sound?

YOU ARE AMAZING!

Your heart is located in your chest and is about the size of your fist. You can feel your heartbeat by touching the side of your neck.

The "tracks" on the circulatory system railroad are: arteries, arterioles, capillaries, venules and veins.
The "tracks" transport O_2 and CO_2.

The right side of the heart receives blood from the superior and inferior vena cava. It then travels to the right atrium, through the tricuspid valve, into the right ventricle, through the pulmonary valve and out the pulmonary artery toward the lungs.

Oxygen-rich blood returns from the lungs through the pulmonary vein. It then travels to the left atrium, through the mitral valve, into the left ventricle, through the aortic valve and out the aorta toward the body.

Heart contraction is generated by the SA (sinoatrial) node. The signal then goes to the AV (atrioventricular) node. The AV node sends the signal to the purkinje fibers which cause the ventricles to contract, moving blood out of the heart.

When the heart relaxes, blood moves back into the heart. The heartbeat sound is made by the closing of the valves.

NEW VOCABU

Arteries

Arterioles

Capillaries

Venules

Veins

Atria

Atrium

Ventricle

Superior

Inferior

Vena Cava

LARY!

Pulmonary

Aorta

Tricuspid Valve

Mitral Valve

Pulmonary Valve

Aortic Valve

SA (sinoatrial) node

AV (atrioventricular) node

purkinje fibers

You are now a Cardiology expert!

Botany:
Plants, Cells, & Photosynthesis

Book 8 of the
Super Smart Science Series

By:
April
Chloe
Terrazas

Ages 0-100

This SUPER AWESOME Book Belongs to:

_____

You are becoming a scientist!

April Chloe Ferrugard

Some plants are carnivorous and they trap their food. Plants that aren't carnivorous have water from the ground.

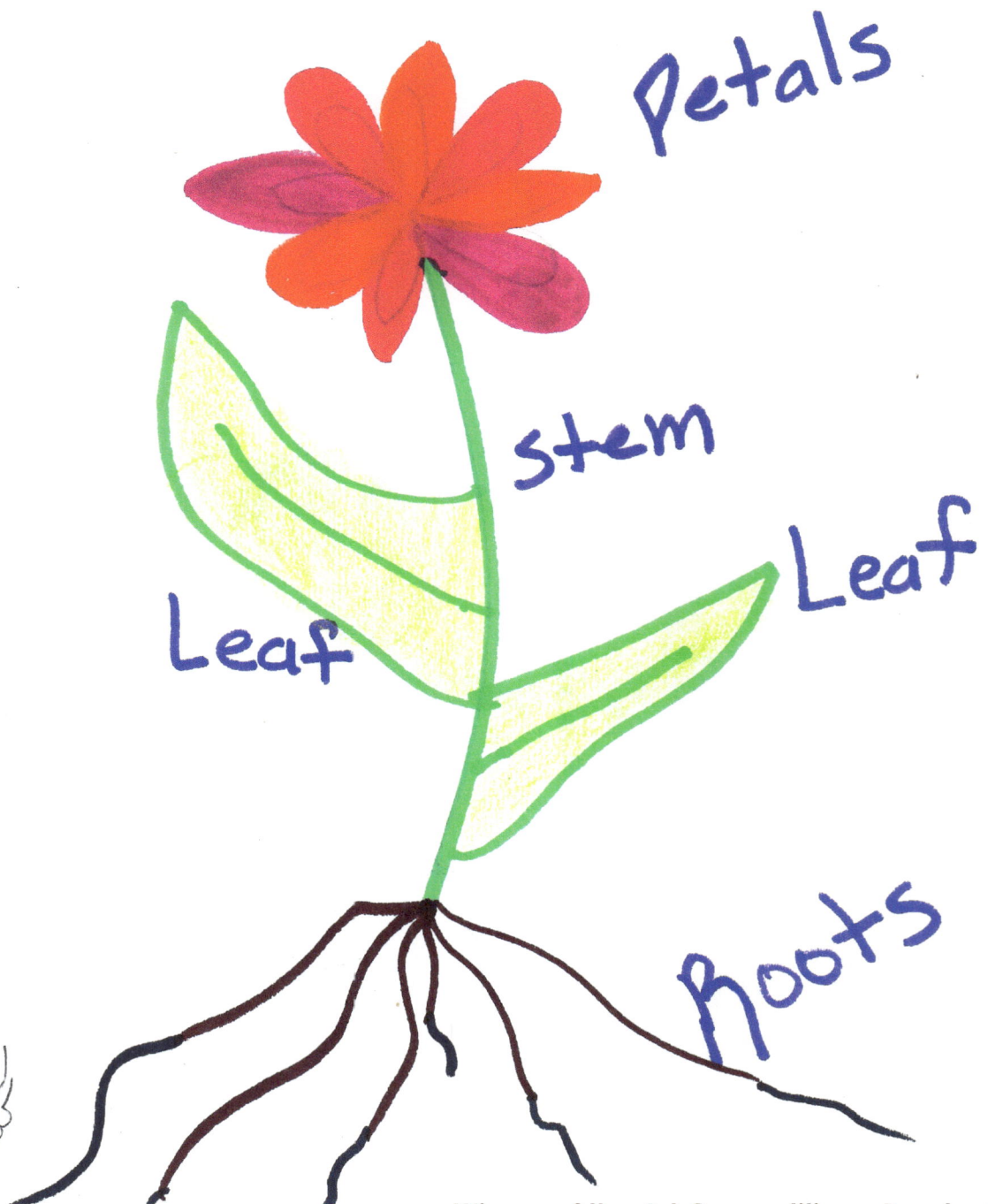

Petals

Stem

Leaf

Leaf

Roots

Brooke Shaw

Winner of the Art Competition - Brooke Shaw!

Botany:
Plants, Cells, & Photosynthesis

This book is dedicated to:
Aunt Katie Ripps

By:
April Chloe Terrazas

Botany: Plants, Cells & Photosynthesis ISBN: 9780984384877
April Chloe Terrazas, BS University of Texas at Austin.
Copyright © 2014 Crazy Brainz, LLC

Visit us on the web! www.Crazy-Brainz.com

Cover design, illustrations and text by: April Chloe Terrazas

This is a plant.

The 5 main parts are:
the root hairs, the roots, the leaf, the stem and terminal bud.

The terminal bud is at
the top of the plant
where new growth happens.

Root hairs are at the bottom
where they can absorb
water and nutrients from the soil.

Just like animals are made of animal cells, plants are made of plant cells.

Terminal Bud →

LEAF

STEM

Root

Root

Root Hairs

Plant Cell

This is a plant cell.

A plant cell has many of the same organelles as an animal cell.

It also has some special parts that are unique only to plant cells.

Cellulose

Sound it Out
1. SEL
2. U
3. LOS

First, look at the cell wall.
The cell wall is made of cellulose.

Plant Cell

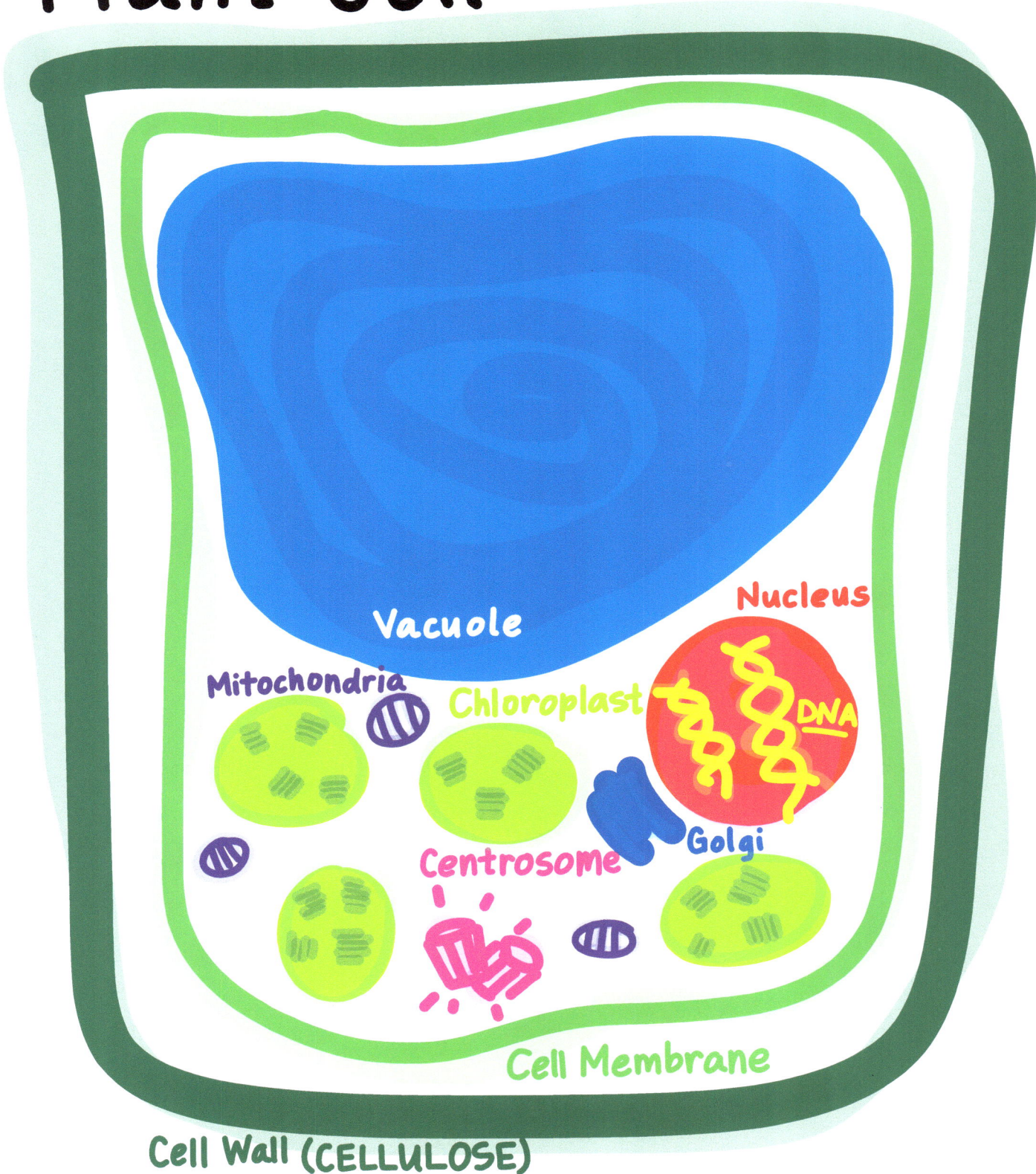

Nucleus

Vacuole

Mitochondria

Chloroplast

DNA

Golgi

Centrosome

Cell Membrane

Cell Wall (CELLULOSE)

Cellulose provides structure and shape for the cell so it can survive in nature.

Cellulose in **cell walls** allows plants to grow very tall!

Does a tall tree have bones like a human?

The strong **cell walls** act like a skeleton and make the tree big and strong.

Cellulose also helps hold up a small plant blowing in the wind.

Vacuole

Vacuoles are very large storage containers in plant cells.

Vacuoles store food, water and nutrients for the plant to survive.

The size of the **vacuole** may increase or decrease depending on how much water the plant has.

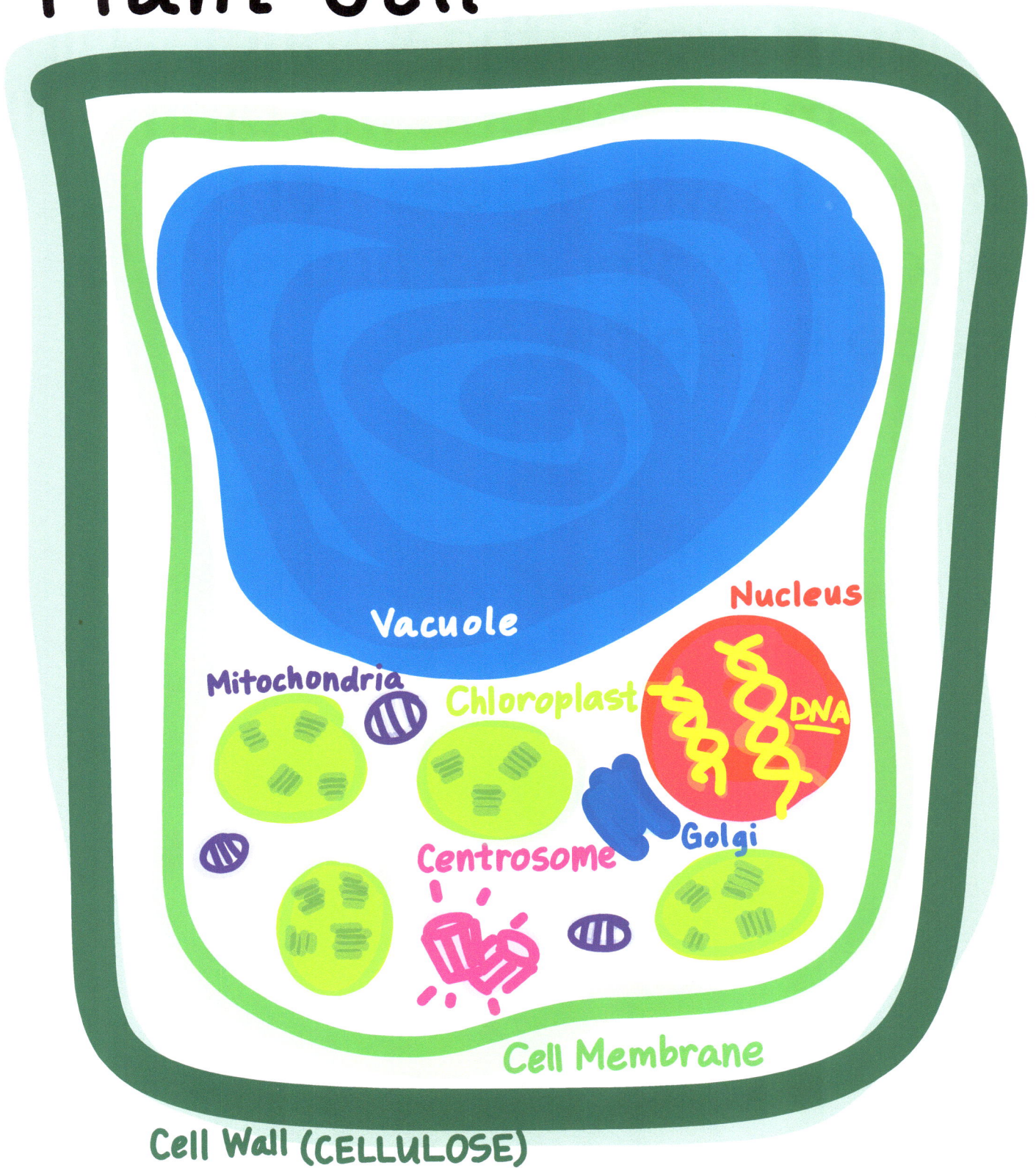

Plant Cell

Nucleus

Vacuole

Mitochondria

Chloroplast

DNA

Golgi

Centrosome

Cell Membrane

Cell Wall (CELLULOSE)

If the plant does not have enough water, the vacuole will become small and the plant will begin to wilt. It does not have enough water and nutrients to keep it healthy.

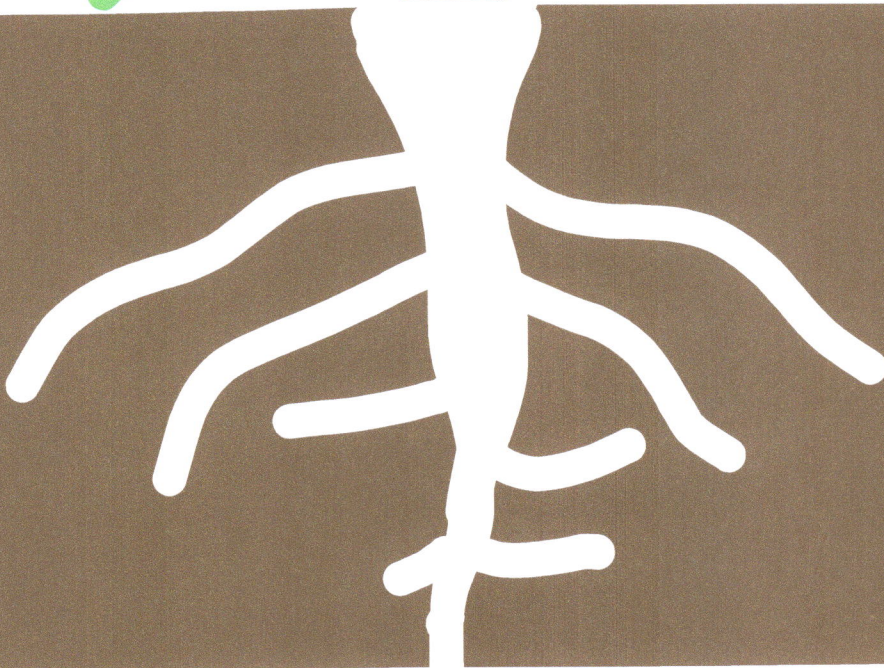

When the plant has plenty of water, the vacuole will be large, full of water and nutrients keeping the plant happy and healthy.

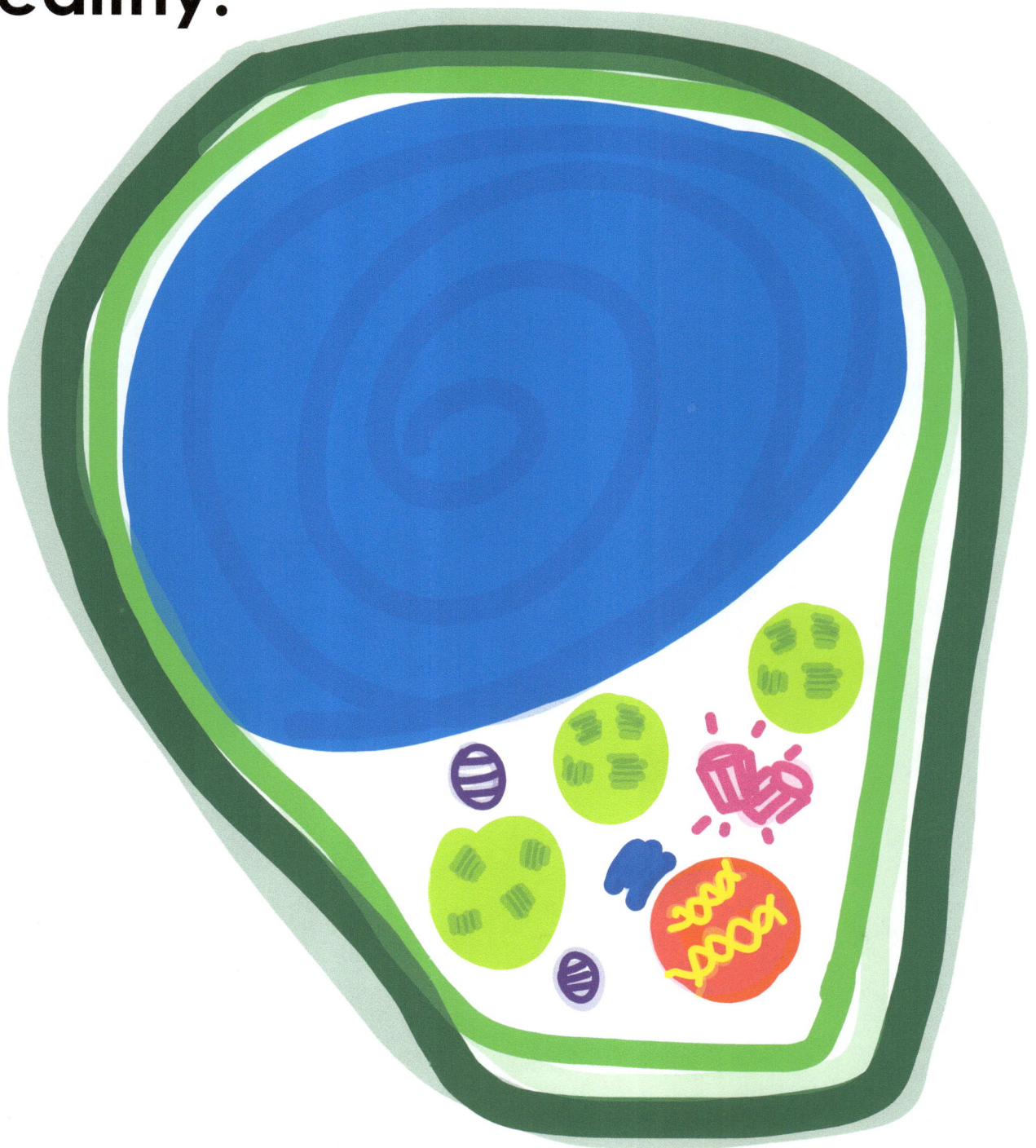

What makes plants green?

The chloroplasts!

Chloroplasts are only found in plant cells.

Chloroplast

Sound it Out

1. **KLOR**
2. **O**
3. **PLAST**

Chloroplasts are like chefs. Chloroplasts use ingredients to make food!

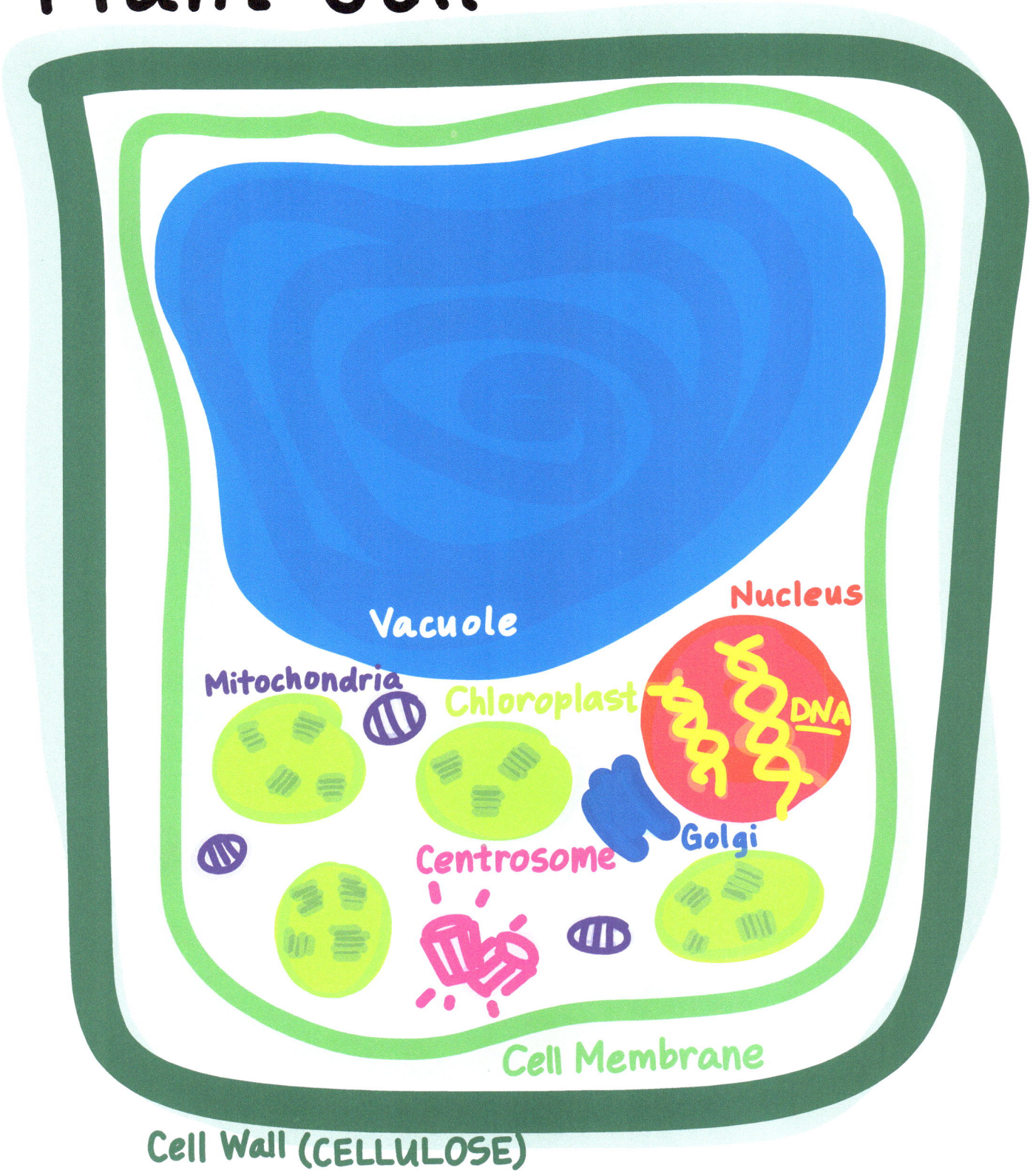

Plant Cell

Vacuole

Nucleus

Mitochondria

Chloroplast

DNA

Centrosome

Golgi

Cell Membrane

Cell Wall (CELLULOSE)

What is cellulose? The vacuole? What are chloroplasts?

Name the parts of a plant:

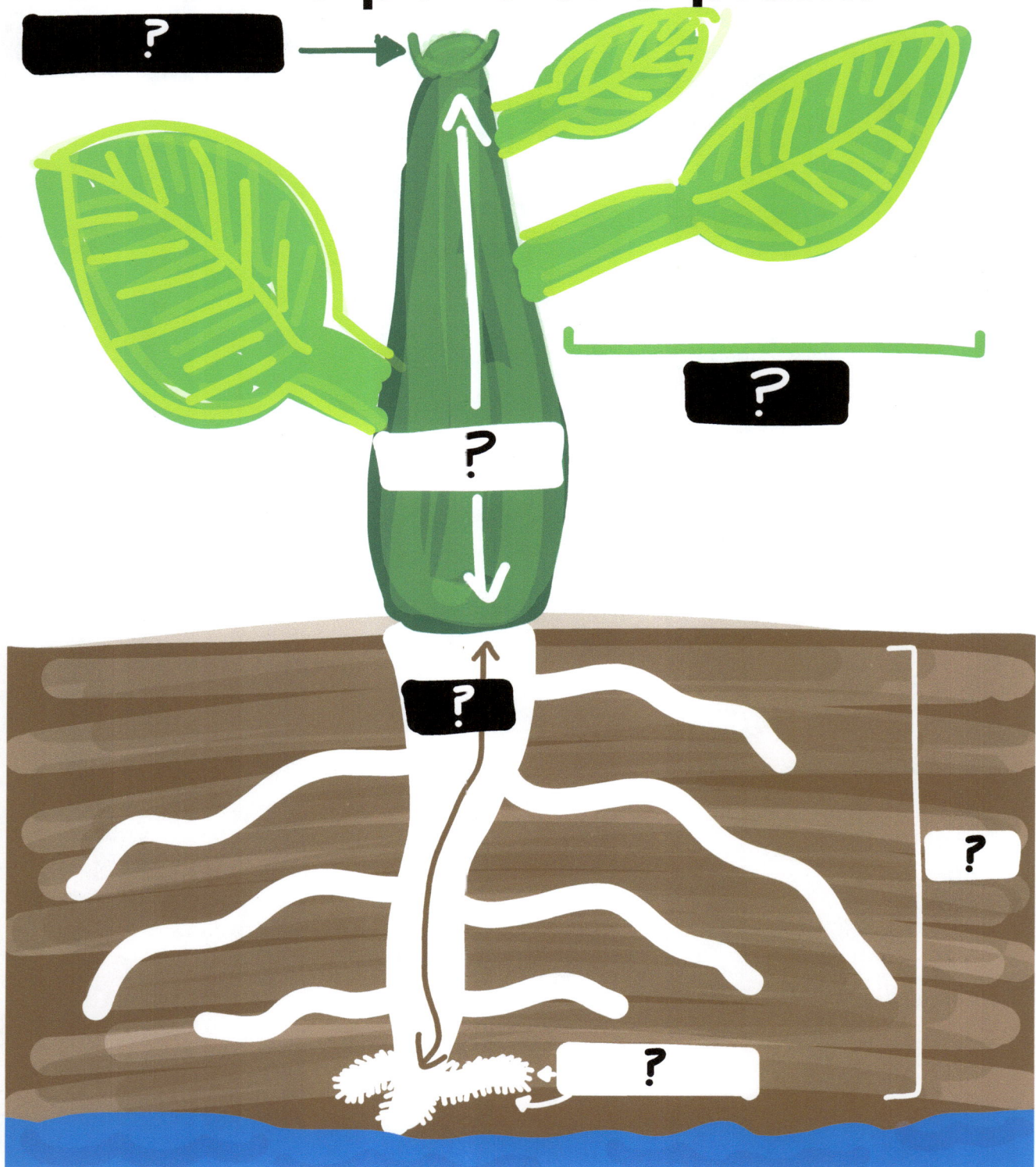

?

?

?

?

?

?

Every green plant that you see is making food using sunlight energy!

Photosynthesis is the process of **chloroplasts** making food using sunlight energy.

Photosynthesis

Photosynthesis is really amazing!

Chloroplasts (in the leaf of the plant) use the following ingredients to make food:

OXYGEN CARBON OXYGEN **+**

Carbon Dioxide
CO_2

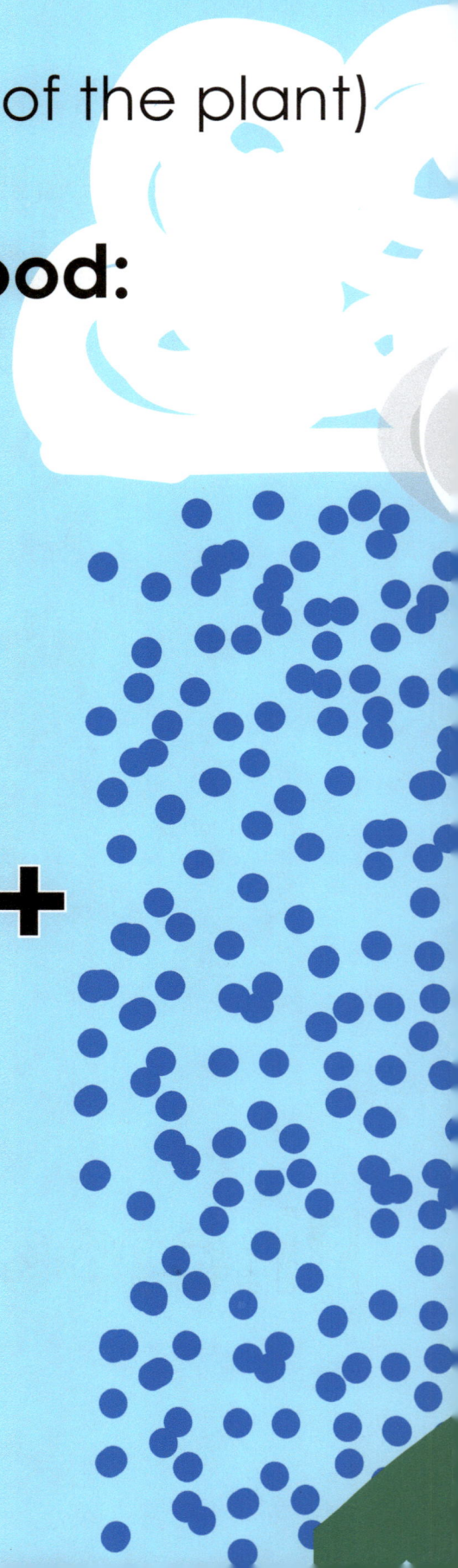

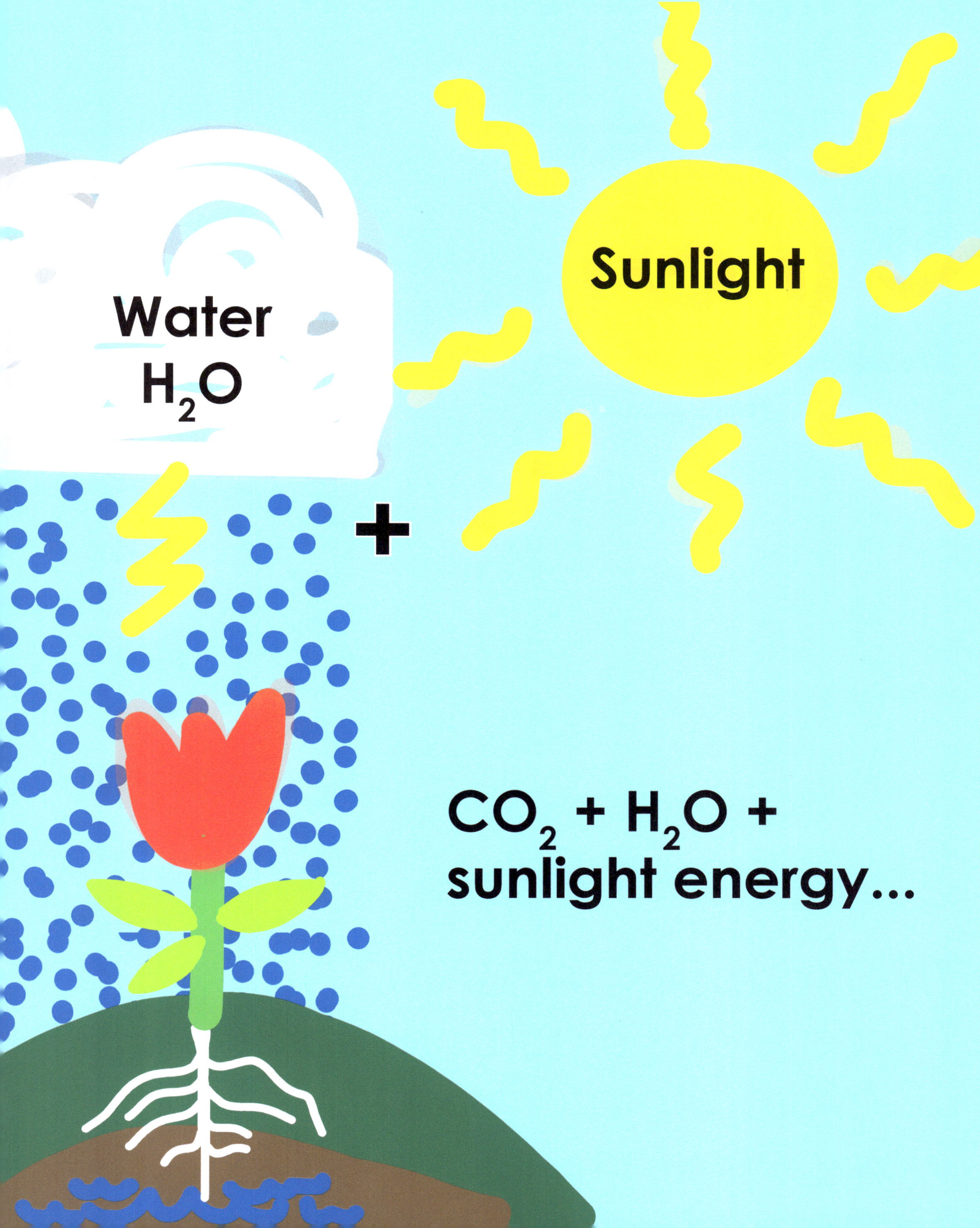

Water
H$_2$O

Sunlight

+

CO$_2$ + H$_2$O +
sunlight energy...

...to make glucose ($C_6H_{12}O_6$)...

Glucose

Sound it Out
1. GLU
2. KOS

$$
\begin{array}{c}
O = C - H \\
H - C - OH \\
HO - C - H \\
H - C - OH \\
H - C - OH \\
CH_2OH
\end{array}
$$

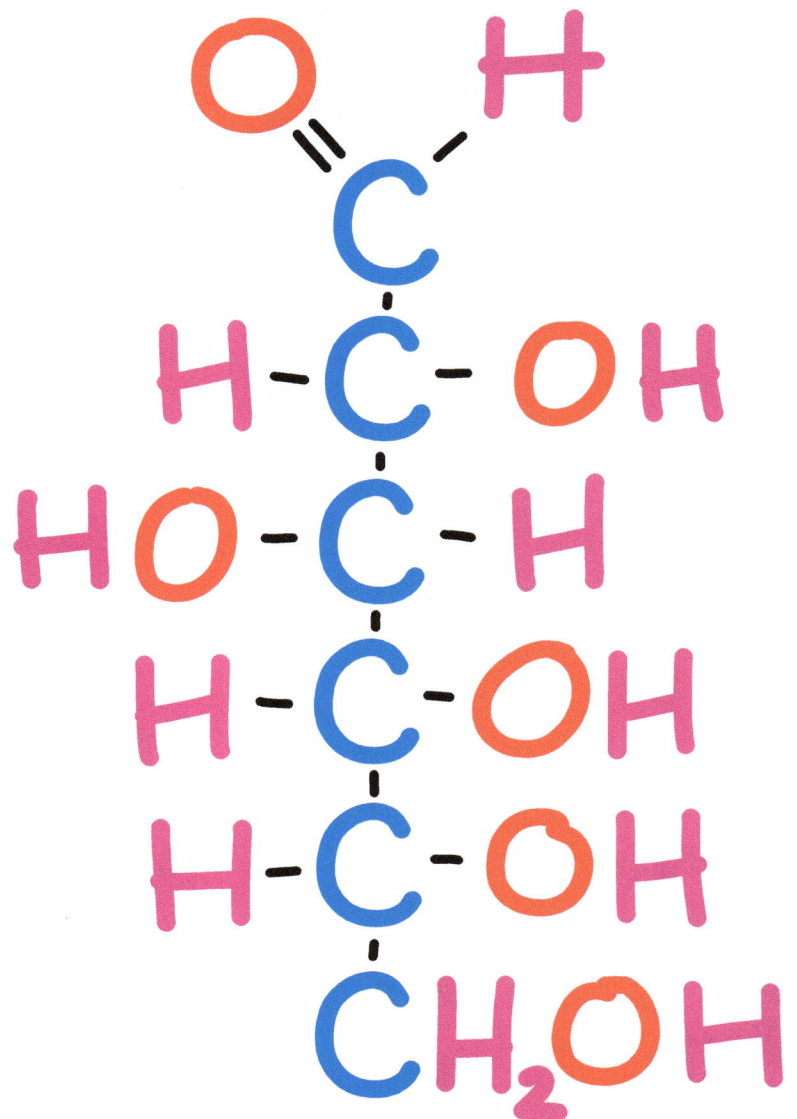

Glucose is a sugar (or food).

This is a molecule of glucose. *Glucose* has 6 carbon atoms, 12 hydrogen atoms and 6 oxygen atoms.

...and oxygen gas (O_2).

The oxygen gas (O_2) that we breathe is made by plants!

+

OXYGEN   OXYGEN

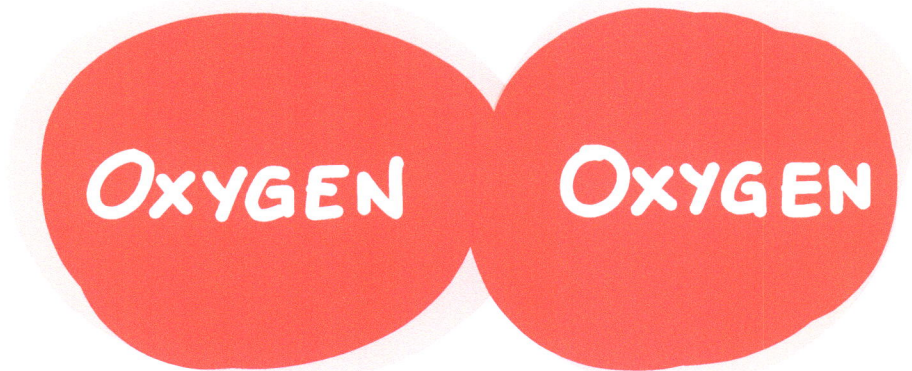

Photosynthesis:

CO_2 + H_2O + sunlight = $C_6H_{12}O_6$ + O_2

Carbon dioxide + water + sunlight = glucose + oxygen gas.

What is photosynthesis?

Where does photosynthesis happen inside the plant cell?

What are the ingredients for photosynthesis ?

What are the products of photosynthesis?

Plants get water from the ground through their roots.

How does water go from the roots to the top of a tree?

Water moves through **xylem** to deliver water to all of the plant cells.

Xylem is like a straw.

Xylem

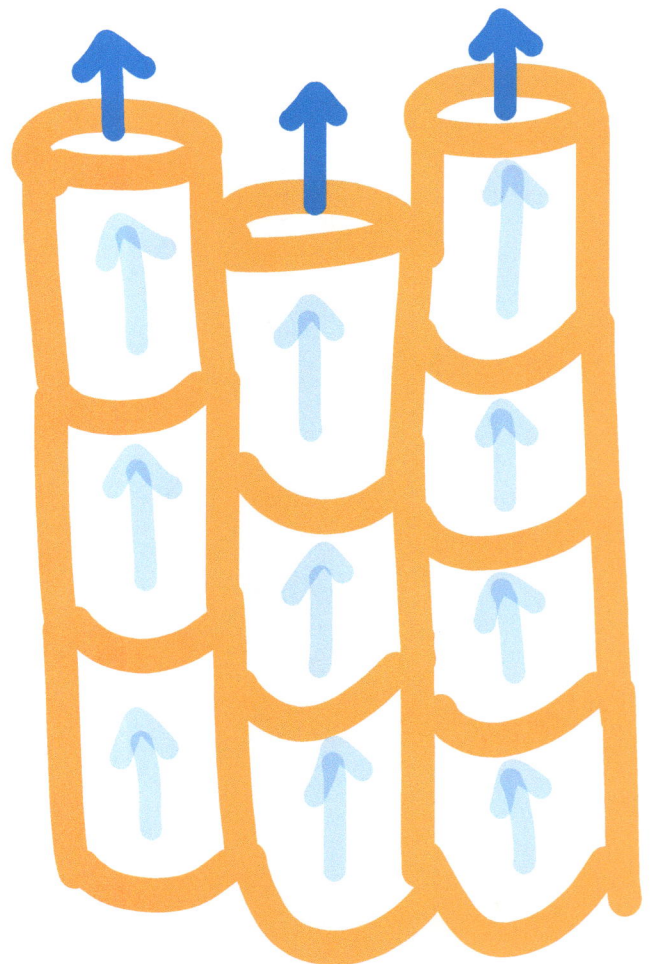

Photosynthesis in the leaves makes sugars (food).

The sugars need to be given to all of the plant cells.

Phloem is another type of straw in the plant that moves sugars from the leaves to all of the cells in the plant, and finally to the roots.

Phloem

Sound it Out

1. FLOM

A plant has 5 main parts: root hairs, roots, stem, leaf, and terminal bud.

The plant cell is different from an animal cell because it has a cell wall made of cellulose.

Inside the plant cell, chloroplasts conduct photosynthesis to make food from ingredients!

The ingredients of photosynthesis are CO_2, H_2O and sunlight.

The products of photosynthesis are glucose and oxygen gas (O_2).

The oxygen gas (O_2) that we breathe is made by plants during photosynthesis.

Glucose is a sugar (food for the plant, and for us!)

Water moves from the ground to the roots, then through the xylem to provide water to all of the cells in the plant.

Sugars (glucose) move from the leaves (where photosynthesis takes place) to all cells in the plant and finally to the roots through the phloem.

Excellent!

<u>*Review your amazing vocabulary:*</u>

Leaf

Stem

Root

Terminal Bud

Root Hairs

Cell Wall

Cellulose

Vacuole

Chloroplast

Photosynthesis

Carbon Dioxide

Oxygen

Hydrogen

Glucose

Xylem

Phloem

You are now a Botany expert!

Marine Biology

April Chloe Terrazas

Book 9 of the Super Smart Science Series™
Ages 0-100

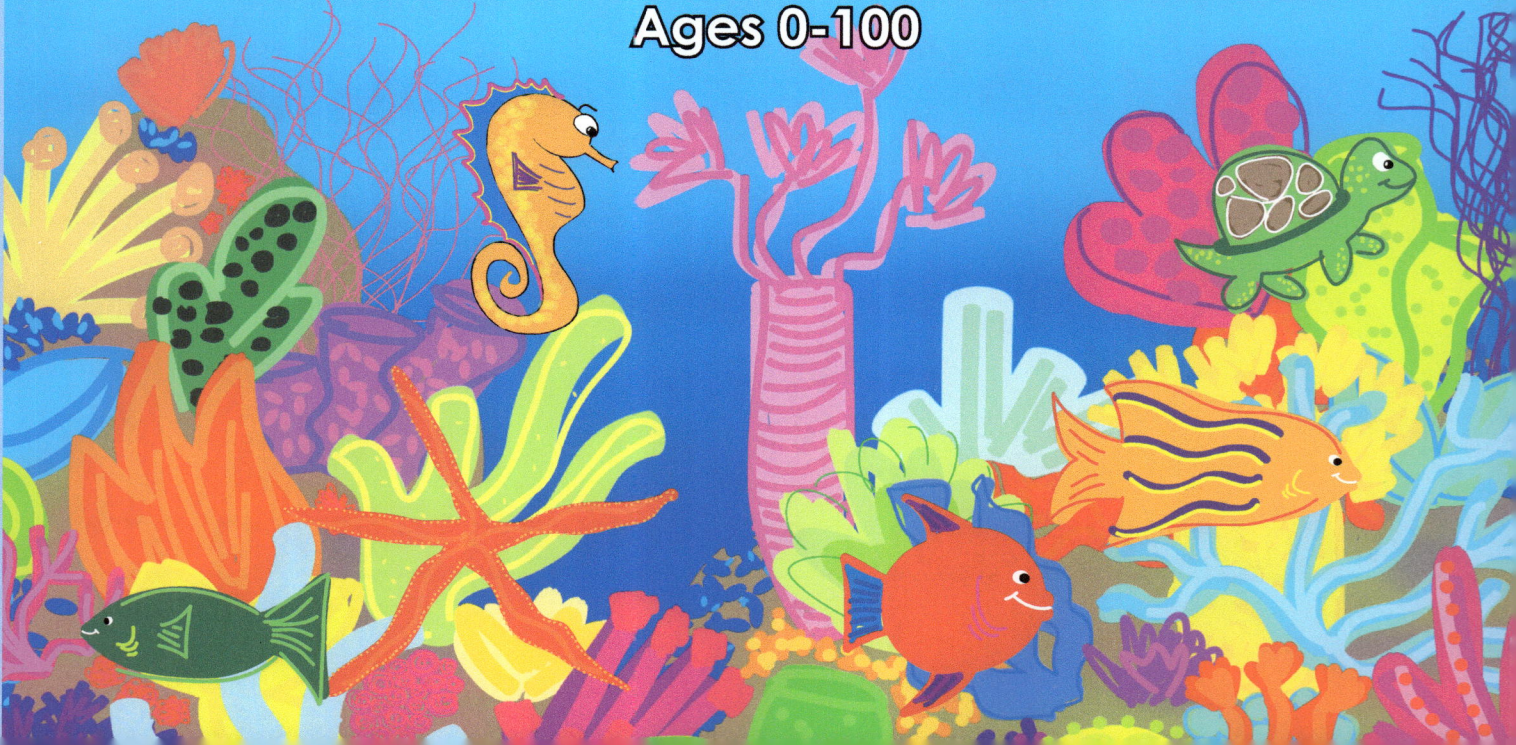

This SUPER AWESOME Book Belongs to:

_____

Winner of the Art Competition: Charlotte Ray

Marine Biology

April Chloe Terrazas

Book 9 of the Super Smart Science Series™
Ages 0-100

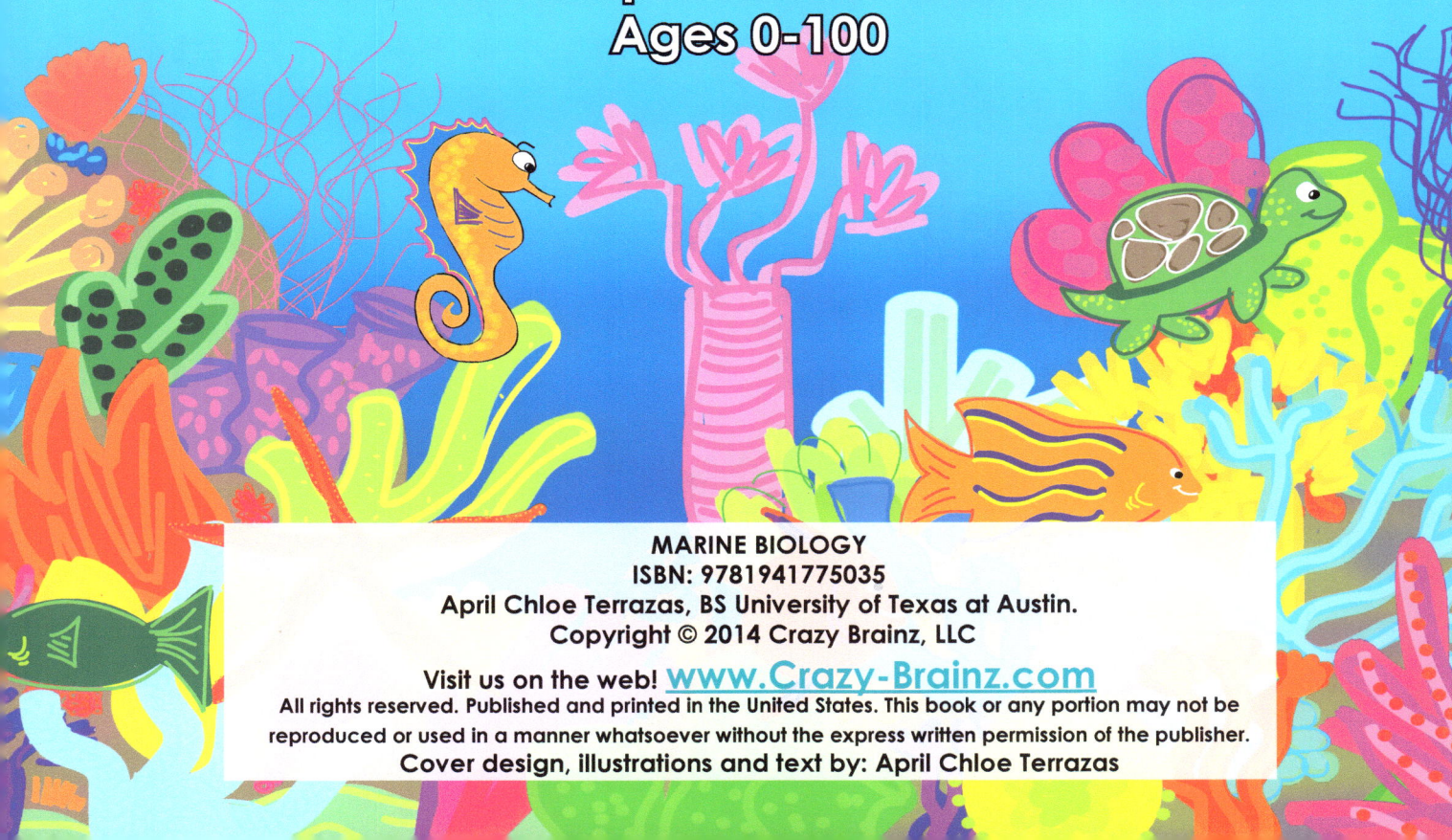

MARINE BIOLOGY
ISBN: 9781941775035
April Chloe Terrazas, BS University of Texas at Austin.
Copyright © 2014 Crazy Brainz, LLC

Visit us on the web! www.Crazy-Brainz.com

Cover design, illustrations and text by: April Chloe Terrazas

Arctic
Ocean

NORTH
AMERICA

Atlantic
Ocean

Pacific
Ocean

SOUTH
AMERICA

Look at all of the
water that
covers Earth!

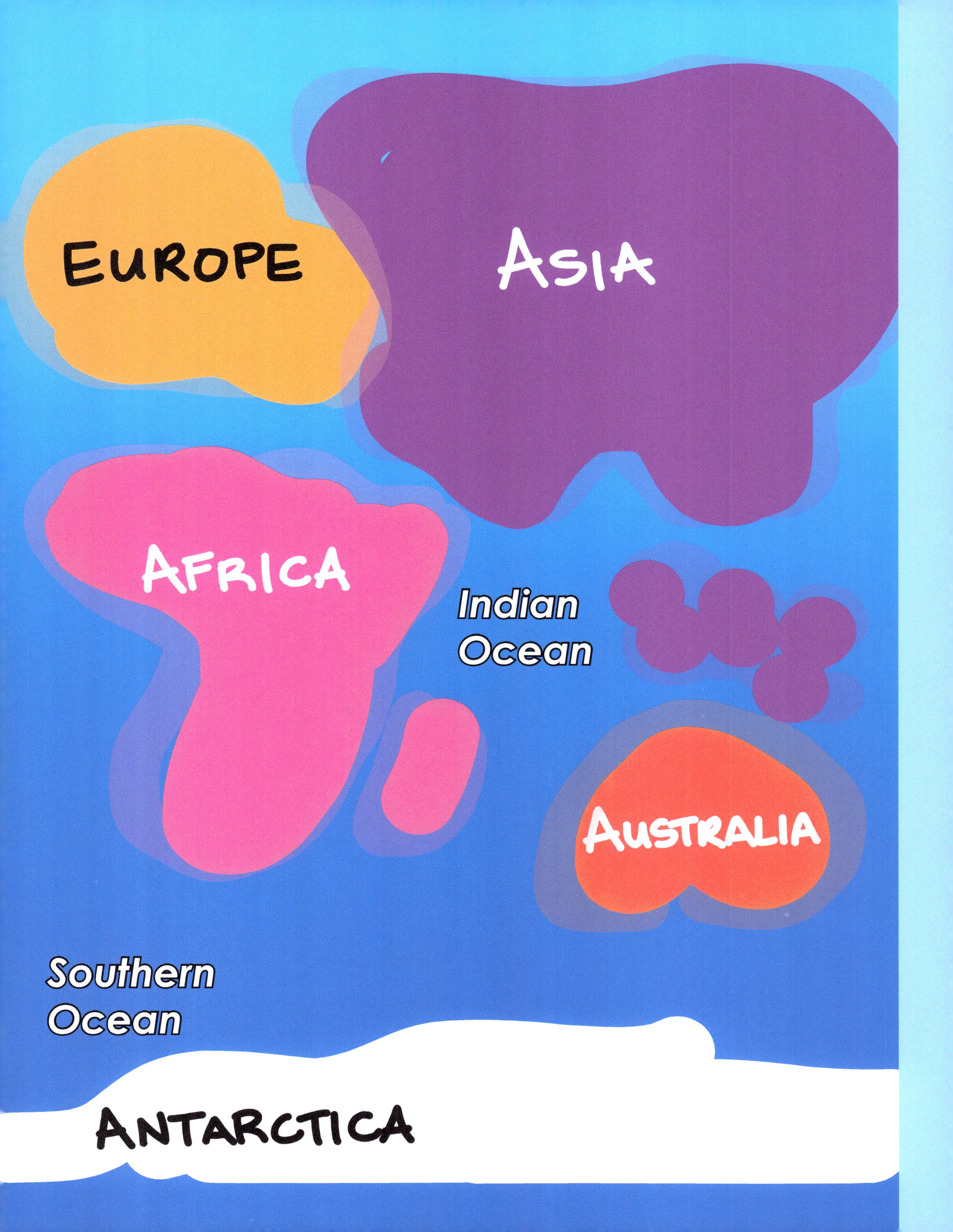
EUROPE
ASIA
AFRICA
Indian Ocean
AUSTRALIA
Southern Ocean
ANTARCTICA

Marine biology means "study of life in the sea."
(Marine = sea,
Biology = study of life)

Over 70% of Earth is covered by water. The oceans are very big, and very deep.

There are many forms of life in the vast oceans on Earth, we have only discovered some of them.

Some sea creatures are very small, some are very big.

You are becoming a
Marine Biologist!

You are going to learn about
sea water composition,
sea plants and sea animals.

Are you ready to begin?

Turn the page
and get started!

Sea water contains Na and Cl.

Na is sodium (SO - DEE - UM).
Cl is chlorine (KLOR - EEN).

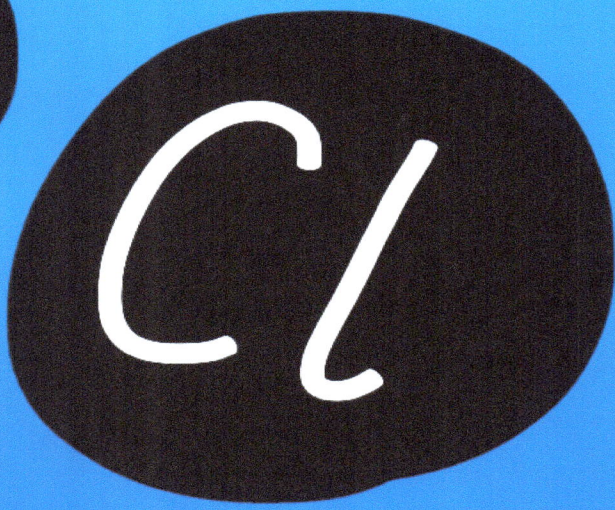

Na and Cl are the main elements that make sea water salty!

PHYTOPLANKTON
(FI - TOE - PLANK - TUN)

Phytoplankton are very small plants that serve as food for many sea creatures.

Phytoplankton are **VERY** small. They are just a tiny dot compared to an ant!

Phytoplankton require sunlight energy for photosynthesis.

PHYTO = plant

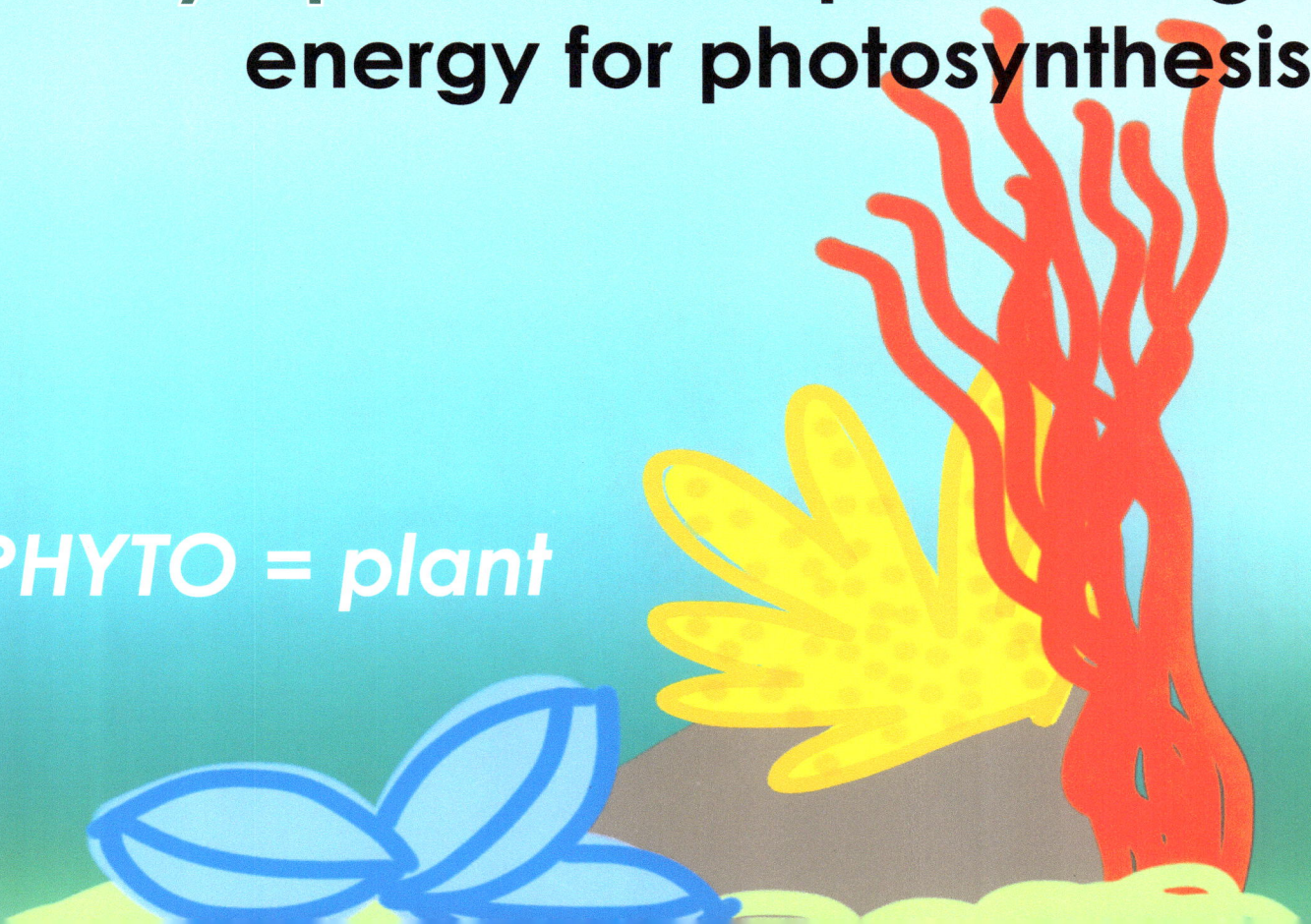

ZOOPLANKTON
(ZOO - PLANK - TUN)

Zooplankton are microscopic (very small) like phytoplankton.
(MI - KRO - SCOP - IK)

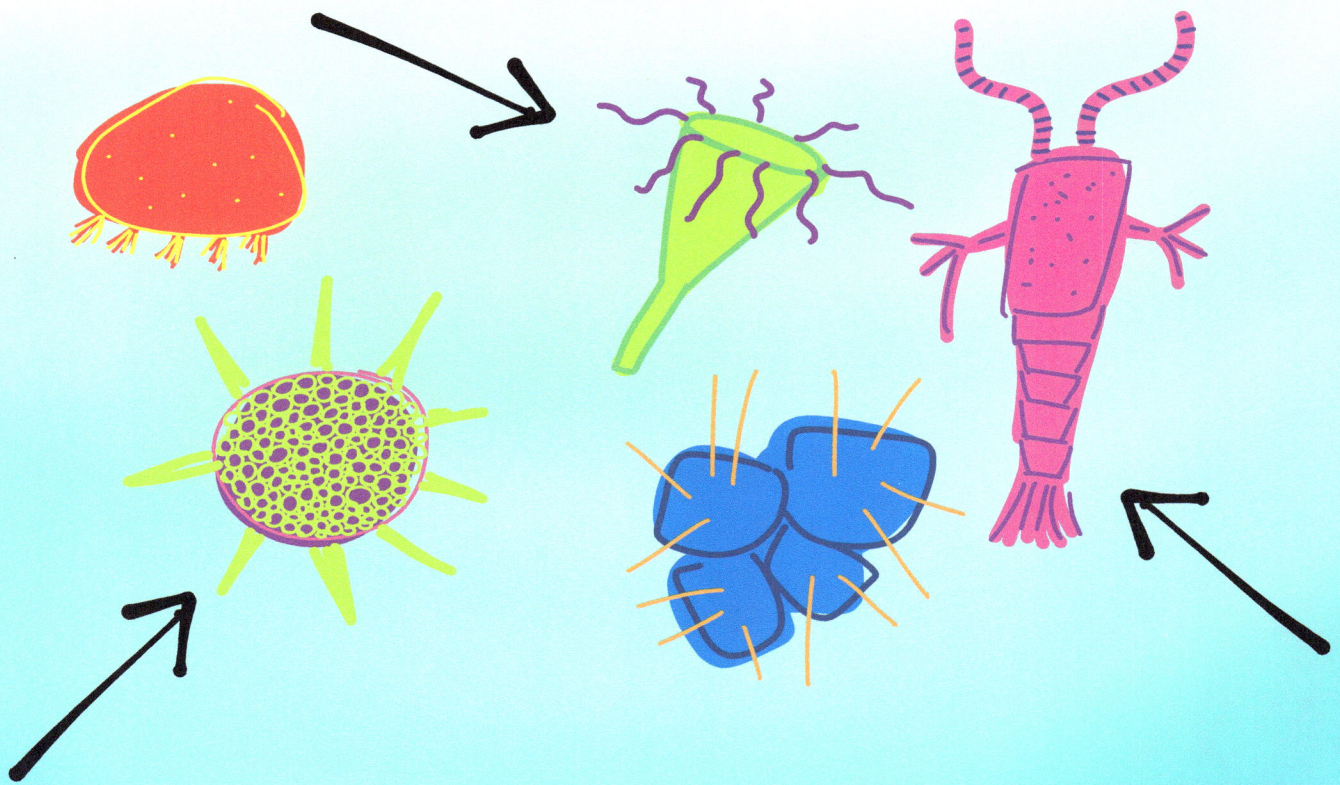

Zooplankton are mostly invertebrate animals.
(IN - VUR - TEH - BRUT)

Zooplankton provide food for a wide variety of marine life.

ZOO = animal

KELP (KELP)

Kelp is a plant.
Kelp provides food and shelter for many sea creatures.

Kelp can grow up to 18 inches in ONE DAY!

Kelp relies on the sun to grow so it is near the surface of the water.

Kelp grows by photosynthesis.

What other sea plants grow by photosynthesis?

What is the definition of "Marine Biology"?

What are the names of the elements that make sea water salty?

What is the name of the very small plants that provide food for many sea creatures?

What are the meaning of the terms "phyto" and "zoo"?

What is the name of the very small invertebrate animals that provide food for many sea creatures?

How many inches can kelp grow in one day?

How do green plants, like kelp, grow?
(hint: it starts with a "P")

Fantastic!
Next, we will learn about animals that live in the ocean.

SPONGE (SPUNJ)

Sponges are sessile. (SES - EL) This means they do not move.

Sponges eat by taking in water through the OSTIA (OS - TEE - UH) *(small holes on the side of the sponge)* and feed on the phytoplankton and zooplankton in the water.

The water and waste then exits the top of the sponge, called the OSCULUM (OS - Q - LUM).

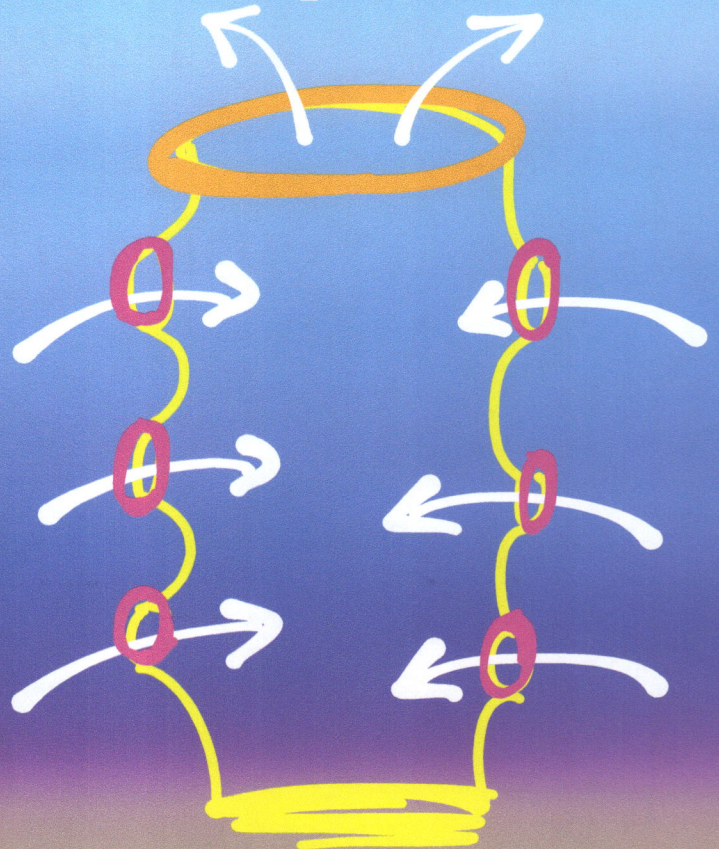

Sponges like to live in warmer, tropical waters.

Sponges do not have a nervous system. A sponge will not respond if you touch it.

SEA STAR (C - STAR)

Sea stars are regenerative.
(RE - JEN - ER - UH - TIV)
This means they can grow a new body from just one piece of their arm!

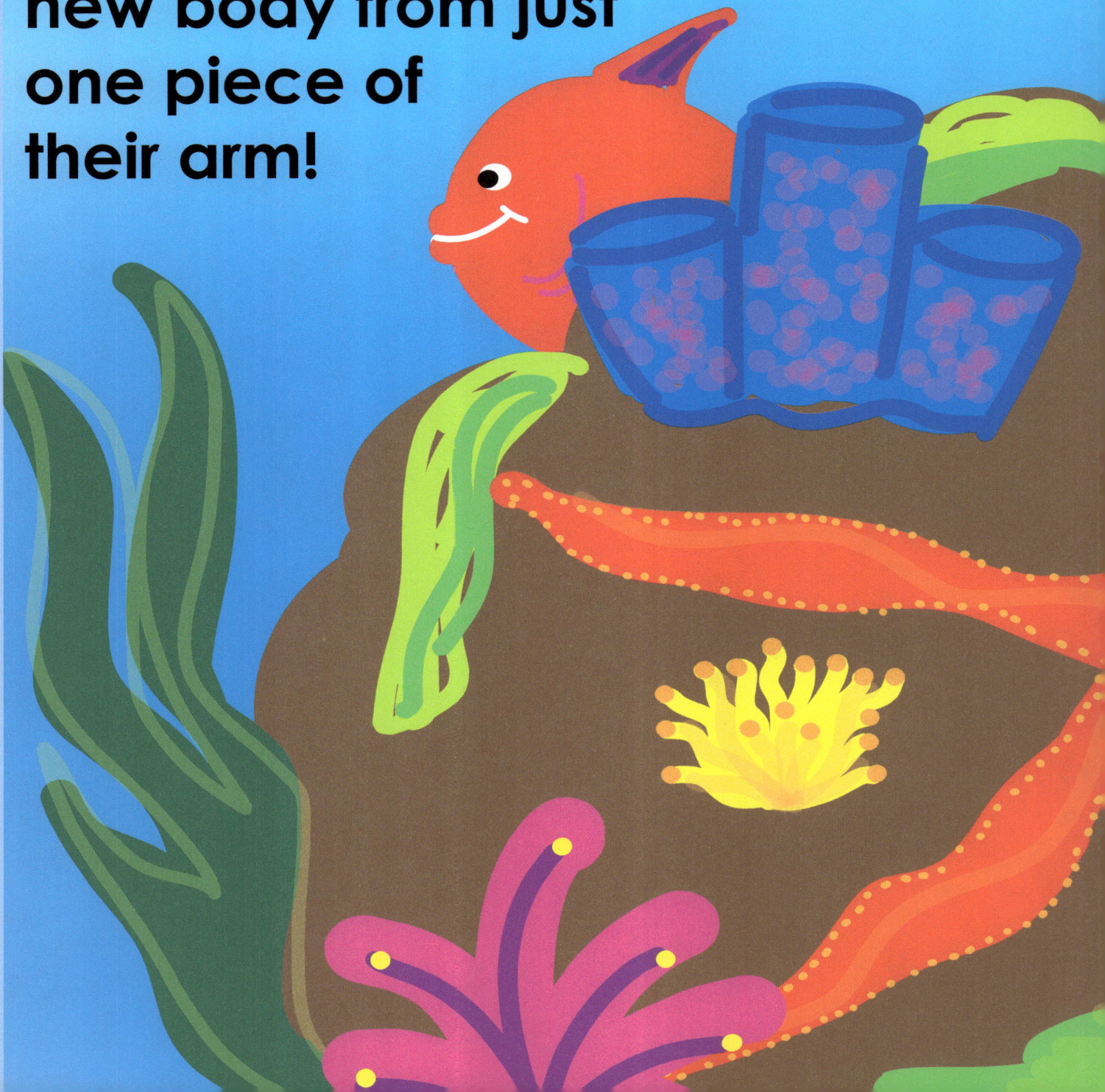

Sea stars come in many sizes and colors.

Sea stars have no brain and no blood. Instead of blood, they have sea water in their circulatory system.

What other marine life do you see here?

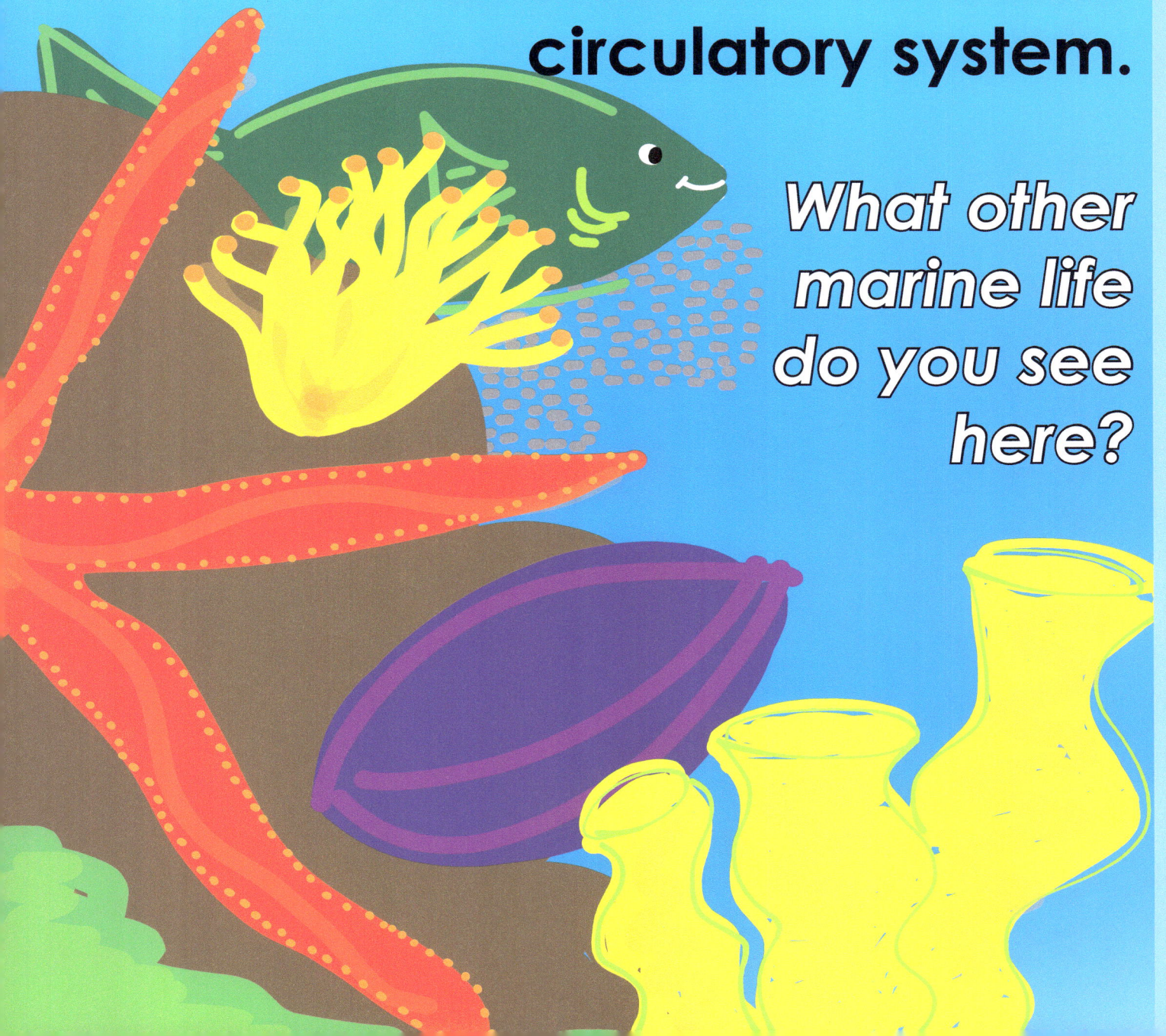

OCTOPUS (OK - TOE - PUS)

OCT = 8

Octopuses live on the sea floor in shallow to very deep waters.

The octopus has a very well developed nervous system which makes it very good at problem solving.

Octopuses secrete a dark ink to distract prey or as a defense to escape a predator.

How many legs does the octopus have?

JELLYFISH (JEL - EE - FISH)

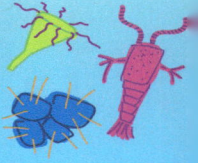

Jellyfish have no bones and no brain. In fact, you can see right through them!

Jellyfish have sensors on their tentacles that detect movement. (TEN - TA - KLZ)

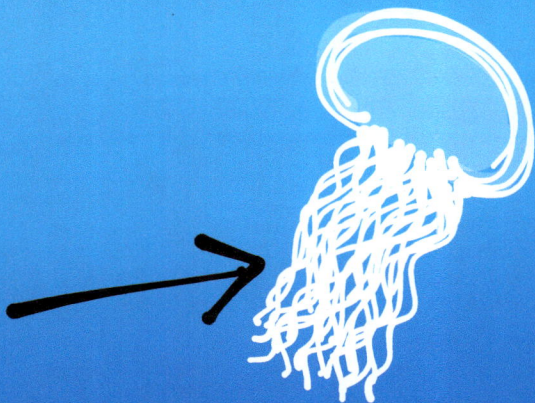

Jellyfish eat phytoplankton, invertebrates and small fish.

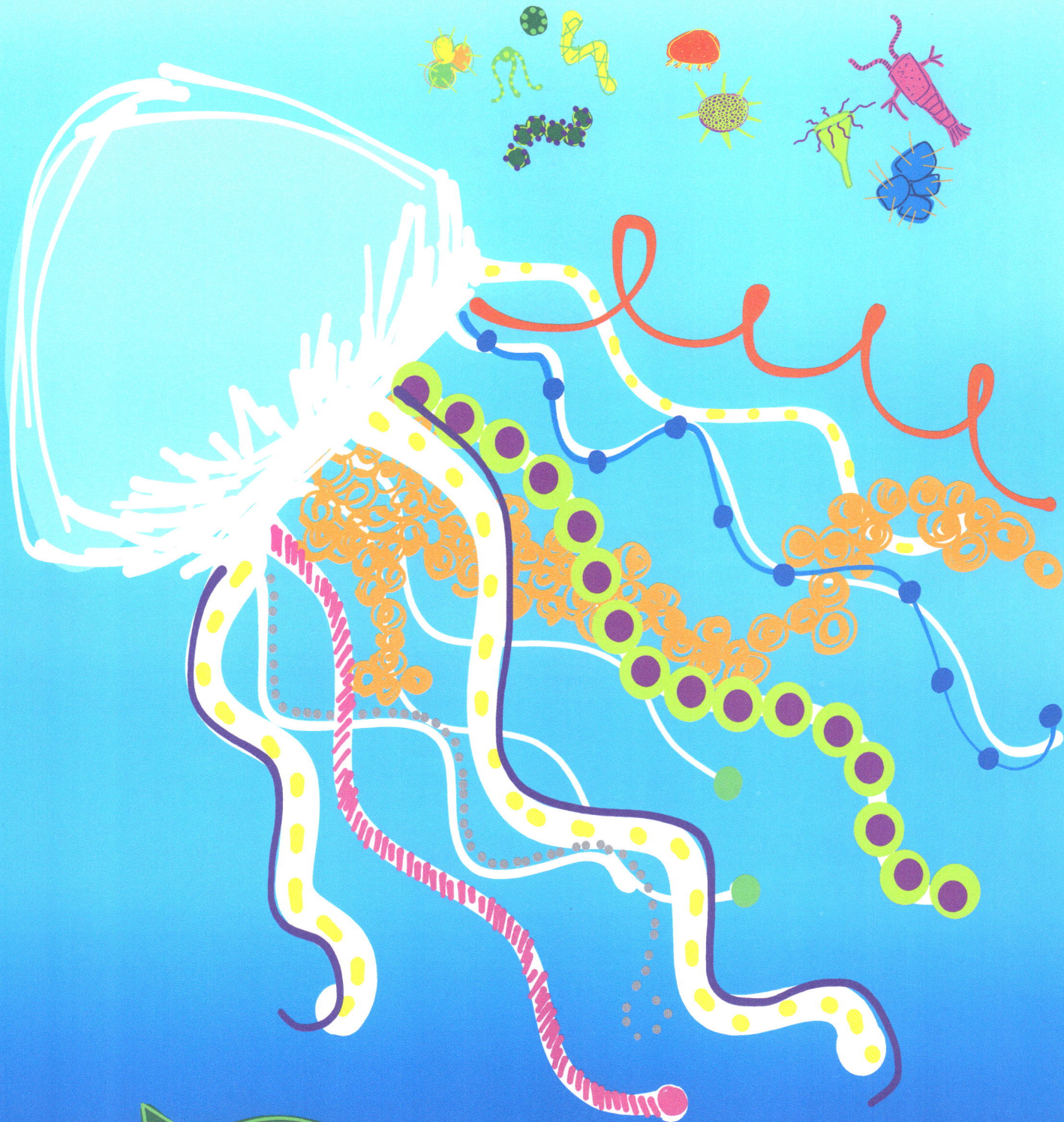

Be careful! Jellyfish have tiny stinging cells in their beautiful tentacles.

Sponges do not move. What is the term meaning "does not move"?

What is the name of the small holes on the sides of the **sponge** that take in water and food?

What is the name of the large hole at the top of the **sponge** where water and waste exit?

Sea stars are regenerative. What does this mean?

TRUE or FALSE

An octopus has 9 legs.

Octopuses secrete a dark ink to distract prey or as a defense against predators.

Octopuses have a poorly developed nervous system.

Jellyfish have no bones, or brain.

Jellyfish will not sting you.

Sea stars have blood in their circulatory system.

Sponges cannot sense touch.

SEAHORSE (C HORS)

Seahorses are very interesting little fish.

Seahorses can move their eyes independently. This means that one eye can be looking left and the other can be looking right!

Seahorses are always hungry. They can eat up to 3,000 pieces of food in just one day!

The seahorse is the only species where the male carries the babies.

Seahorses move by flapping their fins 30 to 70 times per second!

Sea Lion (C LI - UN)

Do you like to swim?
Sea lions LOVE to swim!
Sea lions eat fish.

Sea lions live along coastlines so you can sometimes see them on the beach, laying in the sun.

Sea lions also hear and see very well, even under water!

Stingray (STING - RAY)

Stingrays stay close to the ocean floor. They will bury themselves in the sand to hide from predators.

The eyes are on top of the body. The mouth is on the underbelly!

Stingrays are docile (DOSS - IL), which means they are friendly, and will swim near divers.

Be careful! **Stingrays** have a poisonous tail that can be harmful to humans.

Great White Shark
(GRAT WITE SHARK)

The great white shark is among the largest predatory animals in the ocean. It can grow between 15 to 20 feet long.

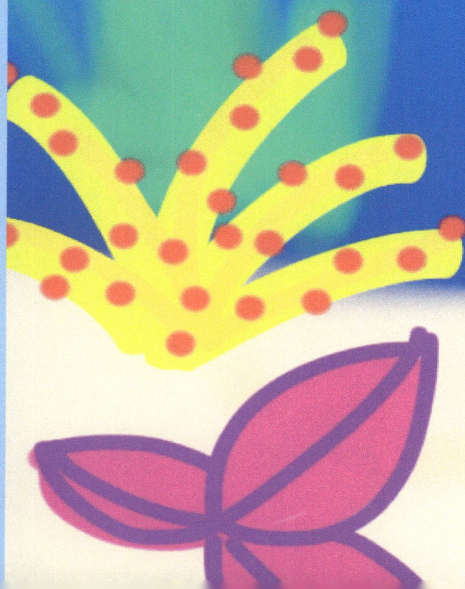

Great white sharks have up to 300 teeth!

Great white sharks can swim very fast, so fast that they can hurl their body completely out of the water!

Great white sharks can weigh up to 5000 pounds.

Coral Reef (KOR-UL REEF)

Corals are animals!
Some corals are hard, making a rigid skeleton of calcium carbonate. ($CaCO_3$, KAL - SEE - UM KAR - BUN - ATE)

Other corals are soft, moving back and forth with the flow of the water.

Coral reefs can be very big.
It takes millions of corals to
make a coral reef.

Coral reefs provide shelter for
many sea creatures.

*What sea creatures do you see
hiding in the coral reef?*

How many pieces of food can a seahorse eat in one day?

Where do stingrays hide from predators?

Where is it possible to see a sea lion?

How much can a great white shark weigh?

What are corals?

Corals make a rigid skeleton of _____ _____.

TRUE or FALSE

Seahorses are the only species where the male carries the babies.

Sea lions do not like to swim.

Stingrays are friendly and will often swim near divers.

Great white sharks have only 30 teeth.

Coral reefs provide shelter for many sea creatures.

Na = Sodium, Cl = Chlorine

Phytoplankton

Zooplankton

Microscopic

Invertebrate

Kelp

Sponge

Sessile

Sea Star

Regenerative

Octopus

Jellyfish

Tentacles

Seahorse

Sea lion

Stingray

Docile

Great White Shark

Coral reef

$CaCO_3$ = Calcium Carbonate

You are now a Marine Biologist!
Next time you see the ocean,
think about all of the
amazing creatures inside!

Ages 0-100

Geology:
Earth Composition, Landforms, Rocks & Water

April Chloe Terrazas

**Super Smart
Science Series
BOOK #10**

Super Smart Science Series Book 10

This SUPER AWESOME Book Belongs to:

_____

Winner of the Art Competition: Daniella Cohen

Geology:
Earth Composition, Landforms, Rocks & Water

Book 10 of the Super Smart Science Series™
Ages 0-100

Geology: Earth Composition, Landforms, Rocks & Water
ISBN: 9781941775066
April Chloe Terrazas, BS University of Texas at Austin.
Copyright © 2014 Crazy Brainz, LLC

Visit us on the web! www.Crazy-Brainz.com

Cover design, illustrations and text by: April Chloe Terrazas

Geology (JEE - OL - O - JEE) **is the study of how the earth was formed and what it is made of.**

Rocks, mountains, rain, snow, and more!

A geologist (JEE - OL - O - JIST) is a person who is an expert in geology.

You will be a geologist soon!

Earth Composition

(KOM - PUH - ZISH - UN)

Earth is made of 4 main layers:

1) **Inner core** (IN - R KOR)
2) **Outer core** (OUT - R KOR)
3) **Mantle** (MAN -TL)
4) **Crust** (KRUST)

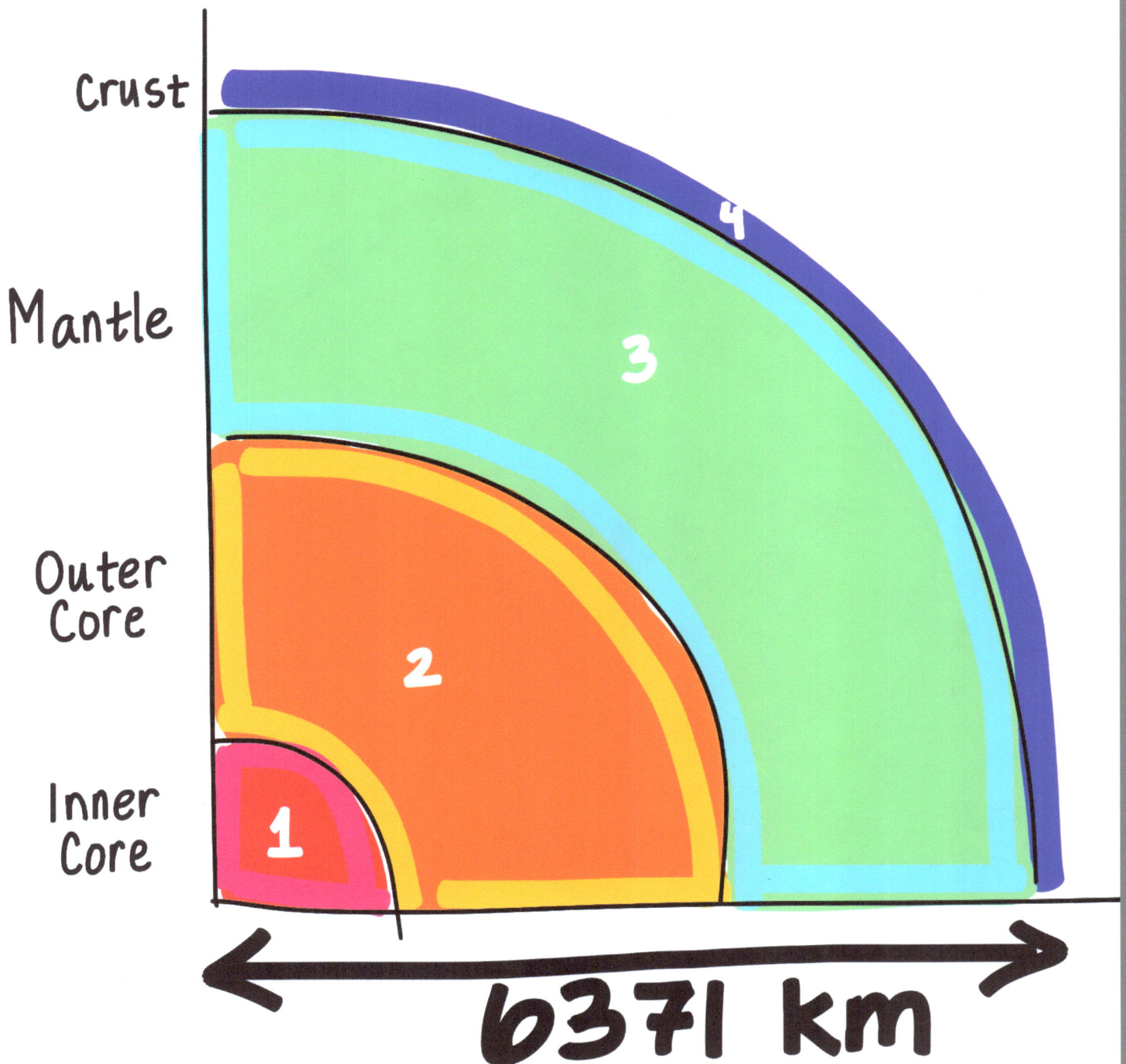

Crust

Mantle

Outer Core

Inner Core

6371 km

The distance from the center of the earth to the surface of the earth is 6,371 kilometers!

(KIL - OM - EH - TRZ)

Driving at 100 kilometers per hour, it would take over 60 hours to reach the center of the earth!

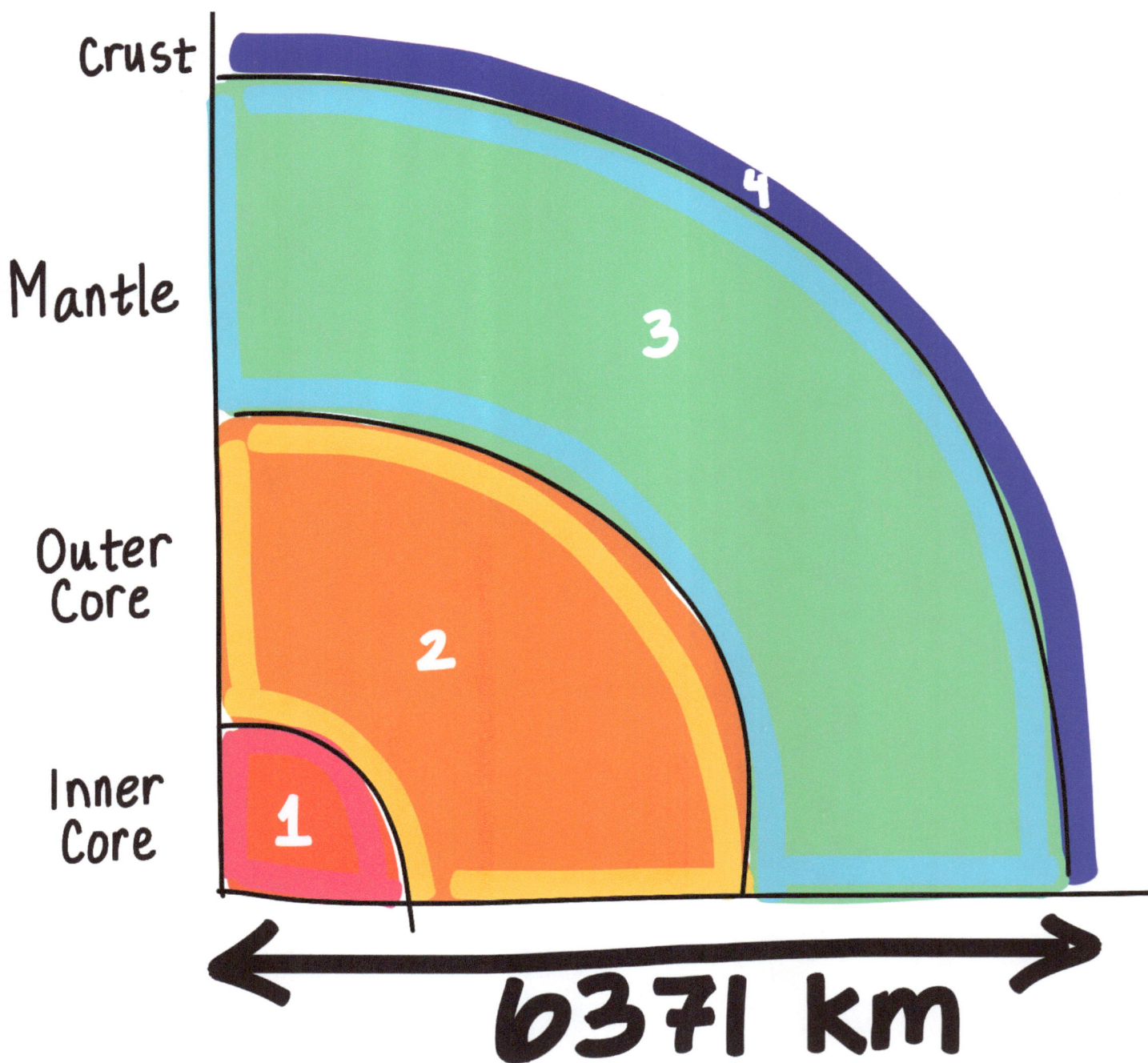

Crust

Mantle

Outer Core

Inner Core

3

4

2

1

6371 km

The core of the earth
(the inner core + outer core)
is made almost entirely of metal.

Temperatures in the inner core
reach 5,000 degrees celsius!
(that's 9,000 degrees Fahrenheit!)

The mantle is the largest part
of the earth and consists
of very hot, solid rock.

The crust is the outer layer
of the earth. It consists of the
continental (KON - TIN - EN - TUL) crust
-covered by land,
and oceanic (O - SHE - AN - IK) crust
-covered by sea.

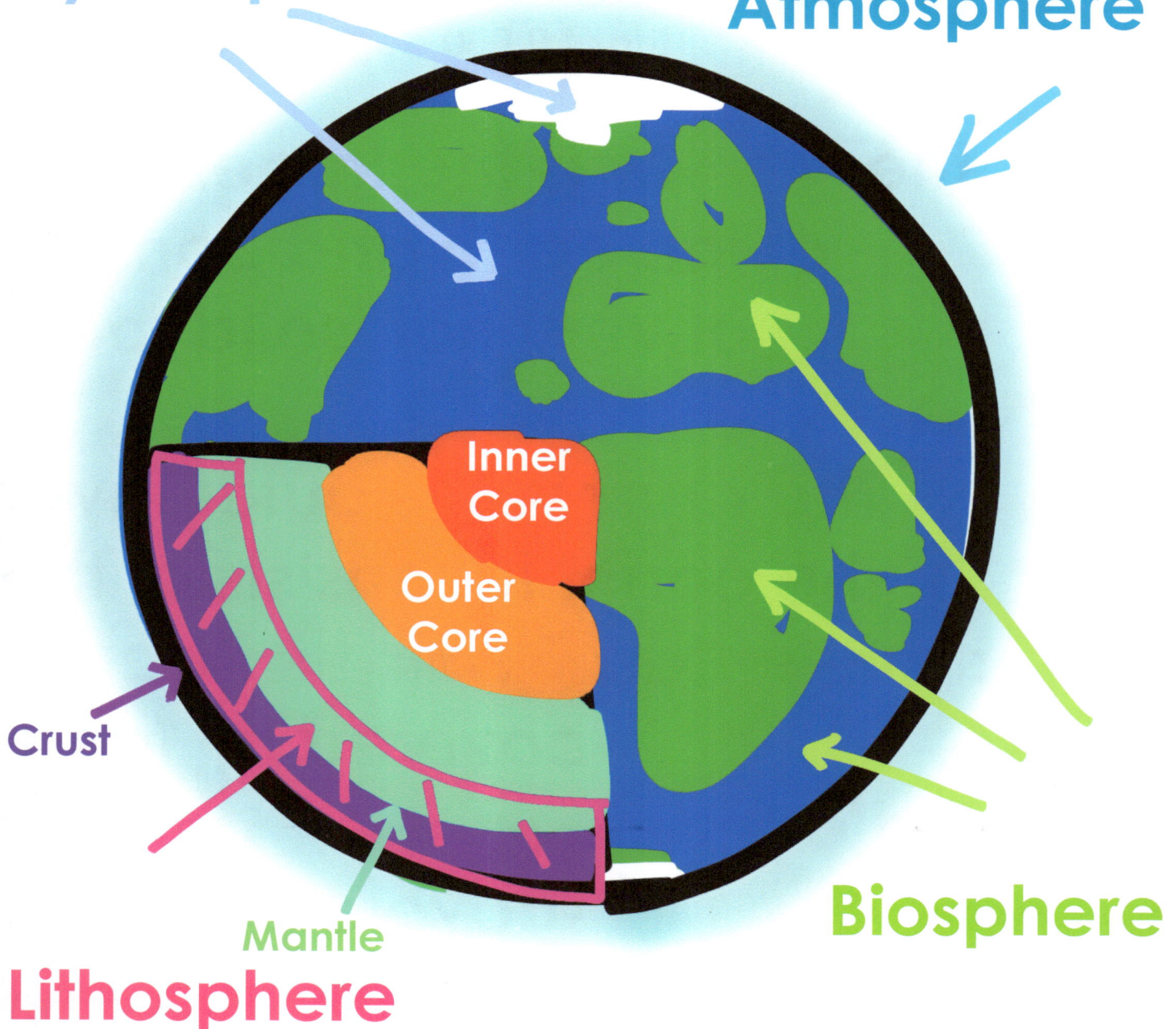

Spheres (SFEERS)

The 4 spheres of the earth interact with each other.

Hydrosphere

Atmosphere

Inner Core

Outer Core

Crust

Mantle

Lithosphere

Biosphere

The hydropshere (HI - DRO - SFEER) is earth's water in solid (ice), liquid (water) and gas (vapor) forms.

The atmosphere (AT - MUH - SFEER) is the layer of gases surrounding our planet. This is the air we breathe! It also protects us from the radiation and heat from the sun.

The biosphere (BI - O - SFEER) is all living things on the earth, on land or in water.

The lithosphere (LITH - O - SFEER) is the rocky layer of the earth's surface. It consists of the crust and a part of the mantle.

Do you remember?

What are the four
main layers of the earth?

Which section is the
largest part of the earth?

What are the names
of the four spheres of the earth?

Excellent!

Tectonic Plates

Tectonic plates (TEK-TON-IK PLAYTS) **are giant pieces of the earth's crust that are part of the lithosphere.**

Plates are slightly moveable.

When plates converge (KUN - VERJ), **one plate will go down below the other, creating earthquakes and causing rock to melt into magma** (MAG - MUH), **which then erupts from volcanoes.** (VOL - KAN - OZ)

This is an example of two oceanic plates converging.

When 2 oceanic plates converge, a very deep trench is formed in the ocean floor and a chain of volcanic (VOL - KAN - IK) islands will form.

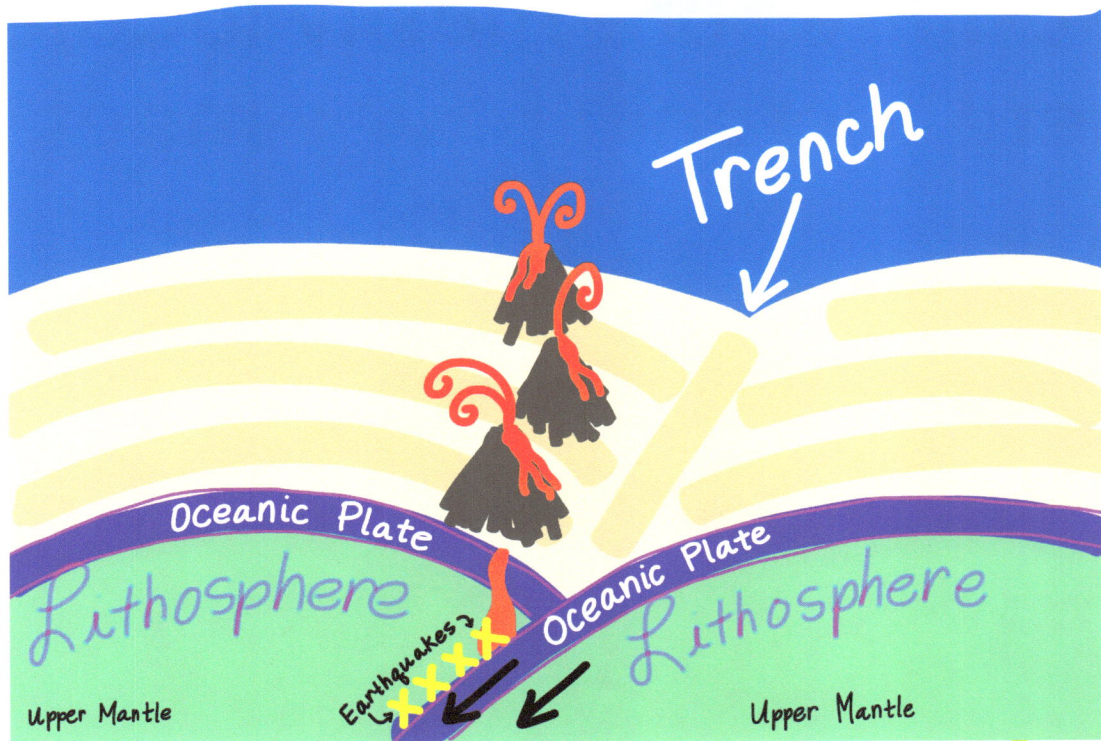

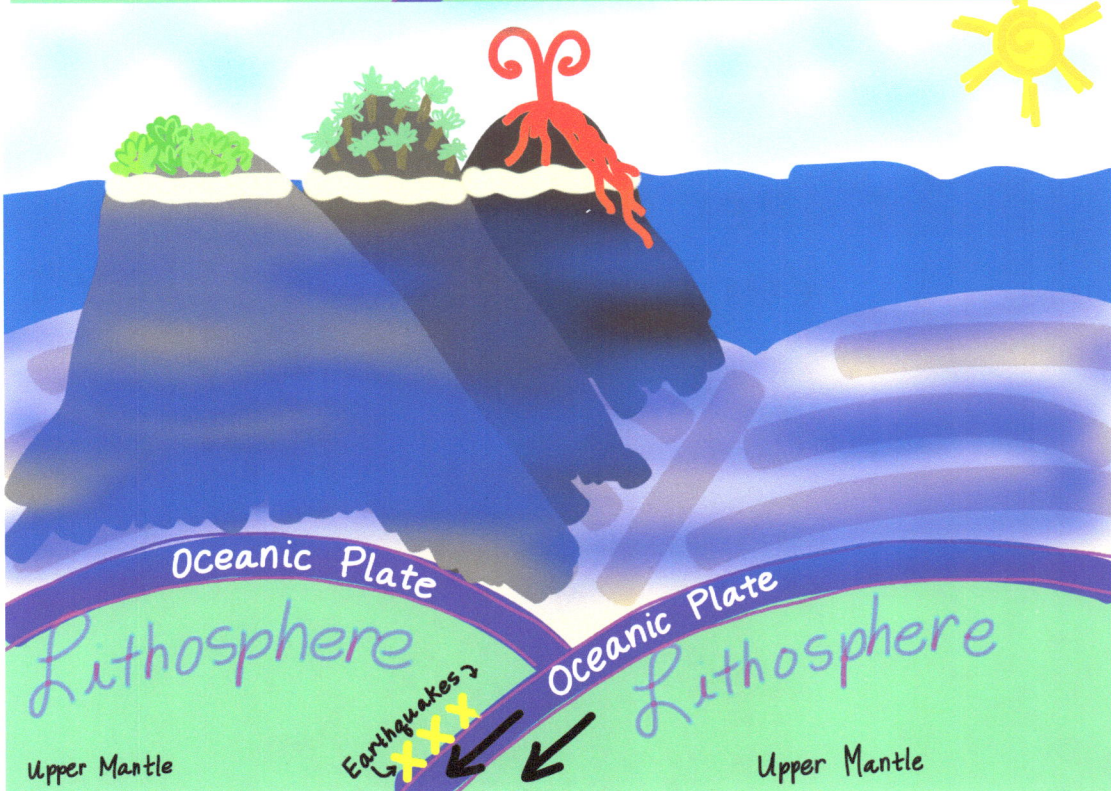

Volcano

Volcanoes are mostly located where tectonic plates meet.

Ash Cloud

Crater
Vent

Lava

Ash

Conduit

Crust

Magma

Mantle

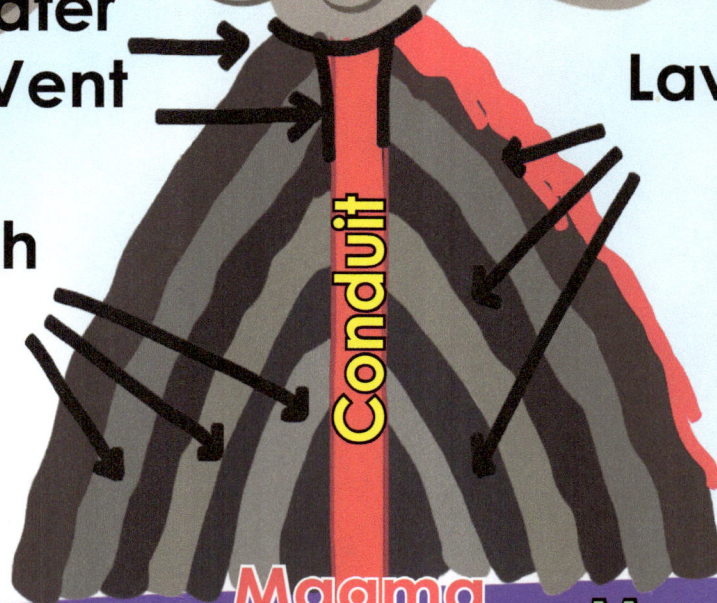

Extremely hot liquid-rock under earth's surface is called magma.

Pressure causes magma to move up through the pipe-like conduit (KON - DU - IT), and out the vent through the crater.

When magma exits the volcano, it is called *lava*.

Lava flows down the surface of the volcano, then ash from the ash cloud layers on top of the *lava*.

This pattern continues as the volcano gets larger.

Landforms

How many of these landforms can you name?

1

2

3

4

5

6

7

8

9

10

11

12

13

14

15

16

17

18

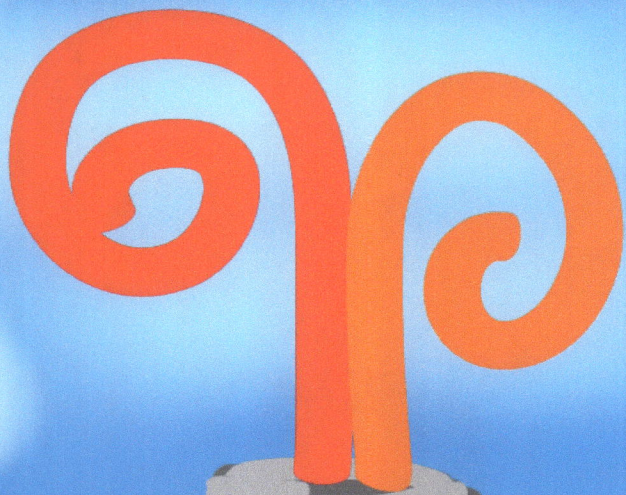

Mountain
MOUN-TUN

Lake
LAYK

Volcano
VOL-KAY-NO

Butte
BYOOT

Plateau
PLA-TOE

Mesa
MAY-SUH

Strait
STRAYT

Harbor
HAR-BR

Islands *I-LUNDZ*

Cape
KAYP

Glacier *GLAY-SHUR*

Hills *HILLZ*

Waterfall *WAH-TR-FALL*

River *RI-VR*

Beach *BEECH*

Delta *DEL-TUH*

Ocean *O-SHUN*

Gulf *GULF*

Do you remember?

How many landforms can you remember?

What is magma?

What "pipe" does magma move through inside the volcano?

What two substances layer on top of one another, building up the volcano over time?

Where are volcanoes mostly located?

True or False

Magma is cold.

When two *plates* **converge,** *volcanoes* **can be formed.**

Earthquakes never occur where two *plates* **converge.**

VERY GOOD!

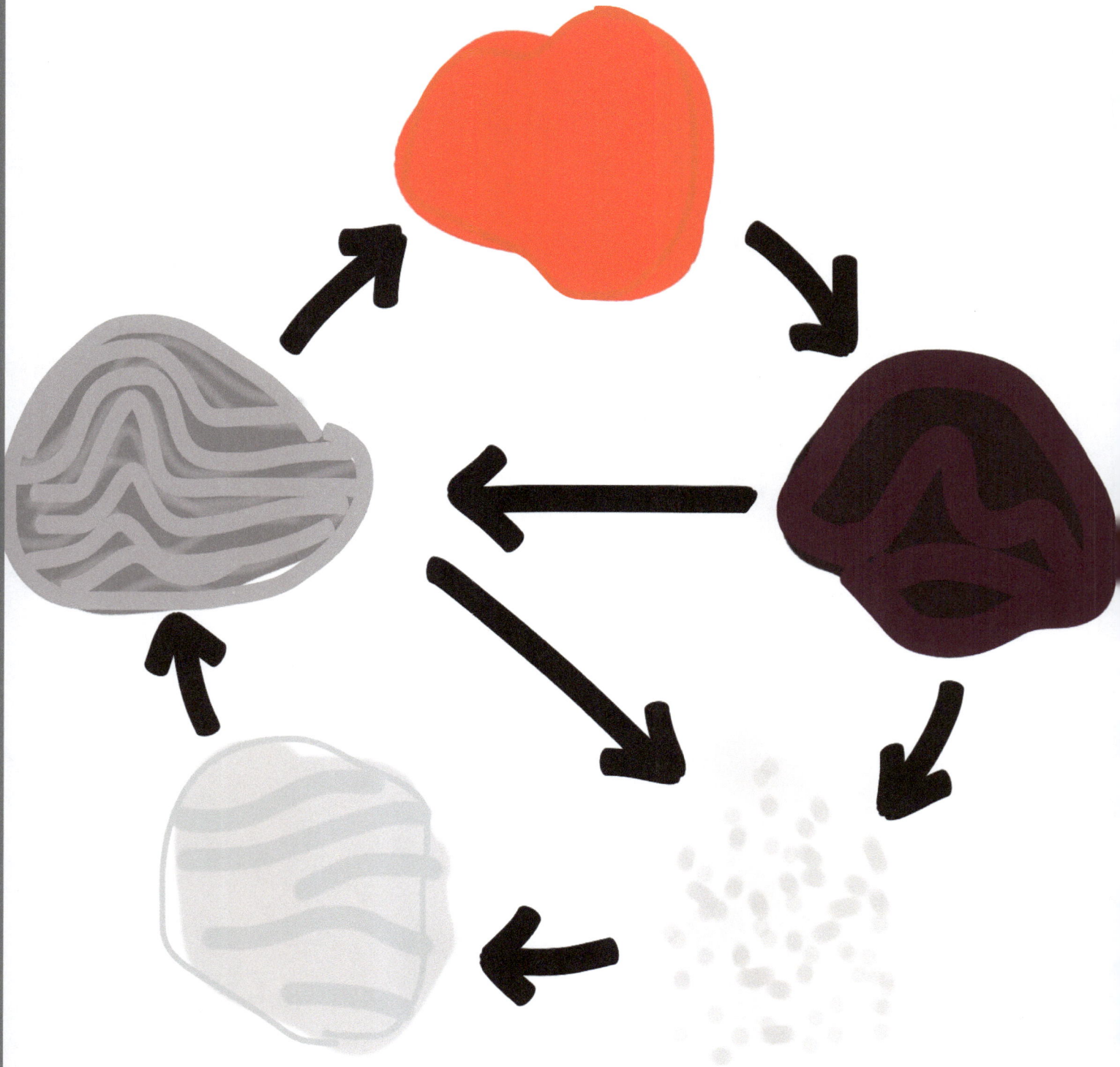

Rock

Cycle

There are three main types of rock on the earth.

Heat, pressure, weathering and erosion change the rocks into different forms.

This is the rock cycle diagram. It shows how the rocks are changed into each form.

There are three main types of rock: igneous, sedimentary and metamorphic.

Igneous rocks are formed when magma cools.

Weathering and erosion break down the rocks into small pieces (sediment). Wind and water move the sediment into piles.

The piles of sediment get covered and buried over time, solidifiying (SUH - LID - EH - FI - ING) and becoming sedimentary rock.

Igneous becomes sedimentary, and igneous can become metamorphic through heat + pressure.

Rock Cycle

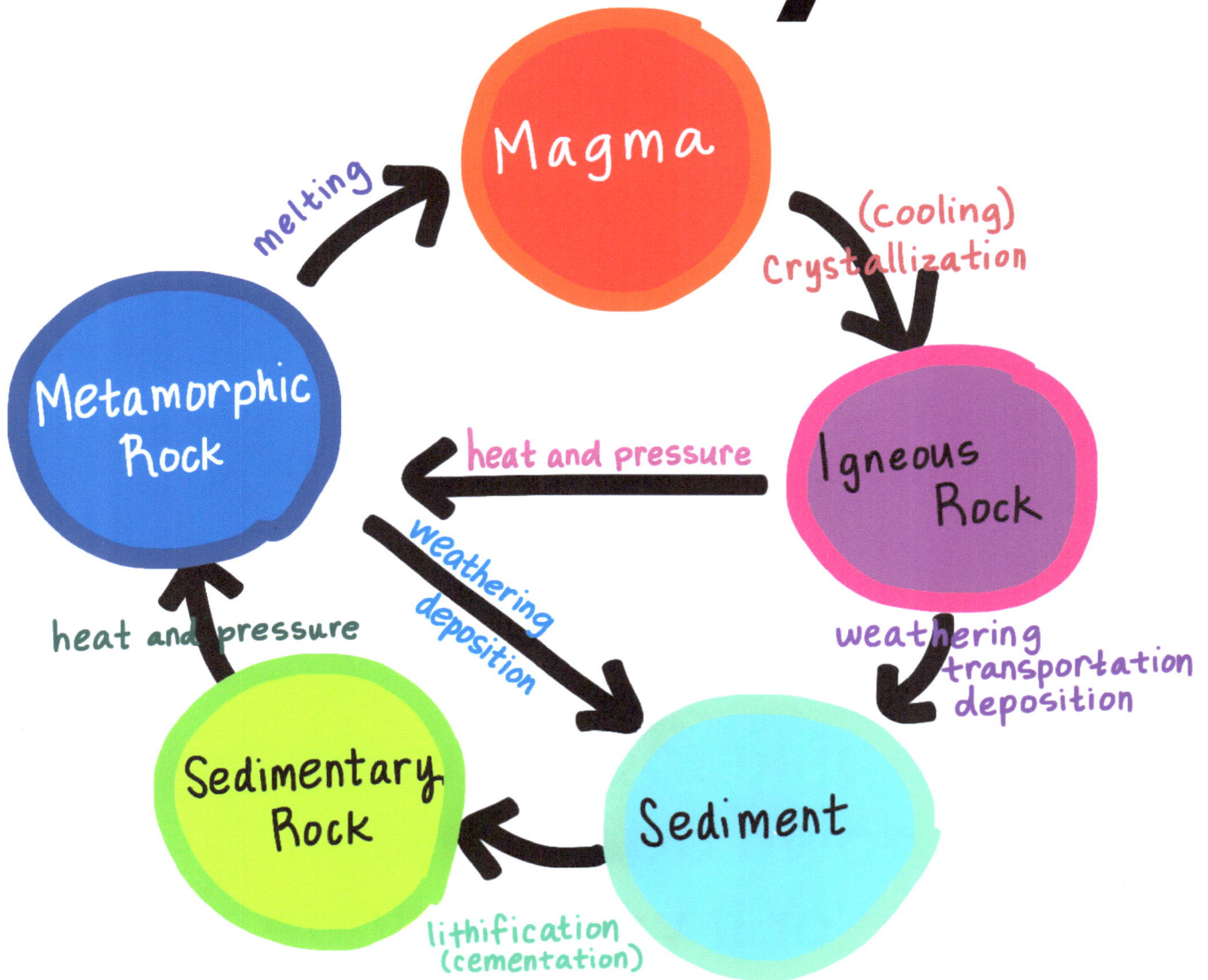

Magma

Metamorphic Rock

Igneous Rock

Sedimentary Rock

Sediment

melting

(cooling) Crystallization

heat and pressure

weathering deposition

weathering transportation deposition

heat and pressure

lithification (cementation)

Igneous: IG - NEE - USS

Sediment: SED - EH - MINT

Sedimentary: SED - EH - MIN - TUH - REE

Metamorphic: MET - UH - MOR - FIK

Sedimentary rock will continue to be buried deeper into the Earth's crust. Heat + pressure changes sedimentary rock into metamorphic rock.

Sedimentary becomes metamorphic.

Metamorphic rock can be broken down again into sediment, or it will continue to be buried deeper into the Earth and melt into magma.

Metamorphic can become sediment or magma.

Rock Cycle

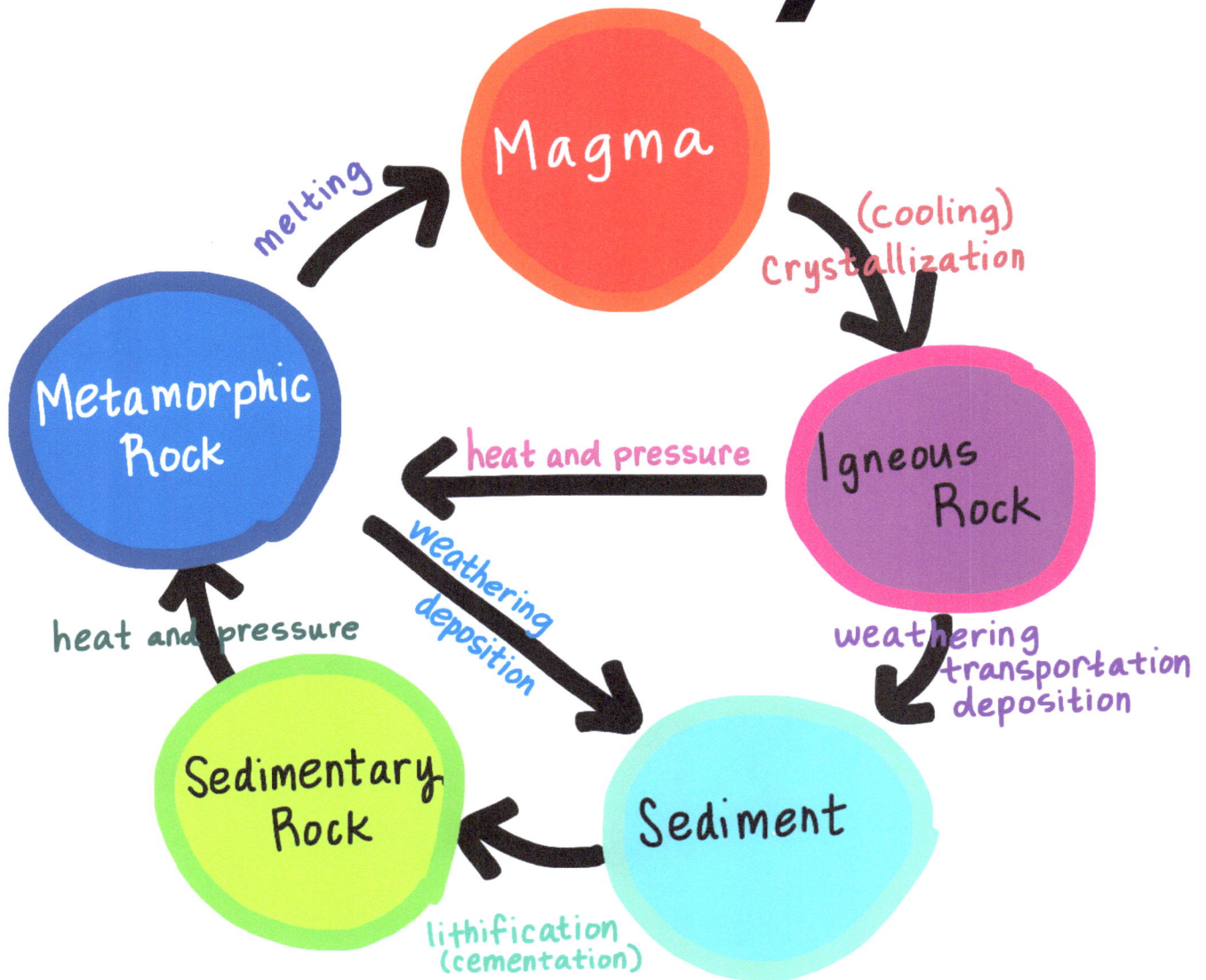

Magma

melting

(cooling)
crystallization

Metamorphic
Rock

heat and pressure

Igneous
Rock

weathering
transportation
deposition

heat and pressure

weathering
deposition

Sedimentary
Rock

Sediment

lithification
(cementation)

Practice saying the name of each part of the rock cycle!

Do you remember?

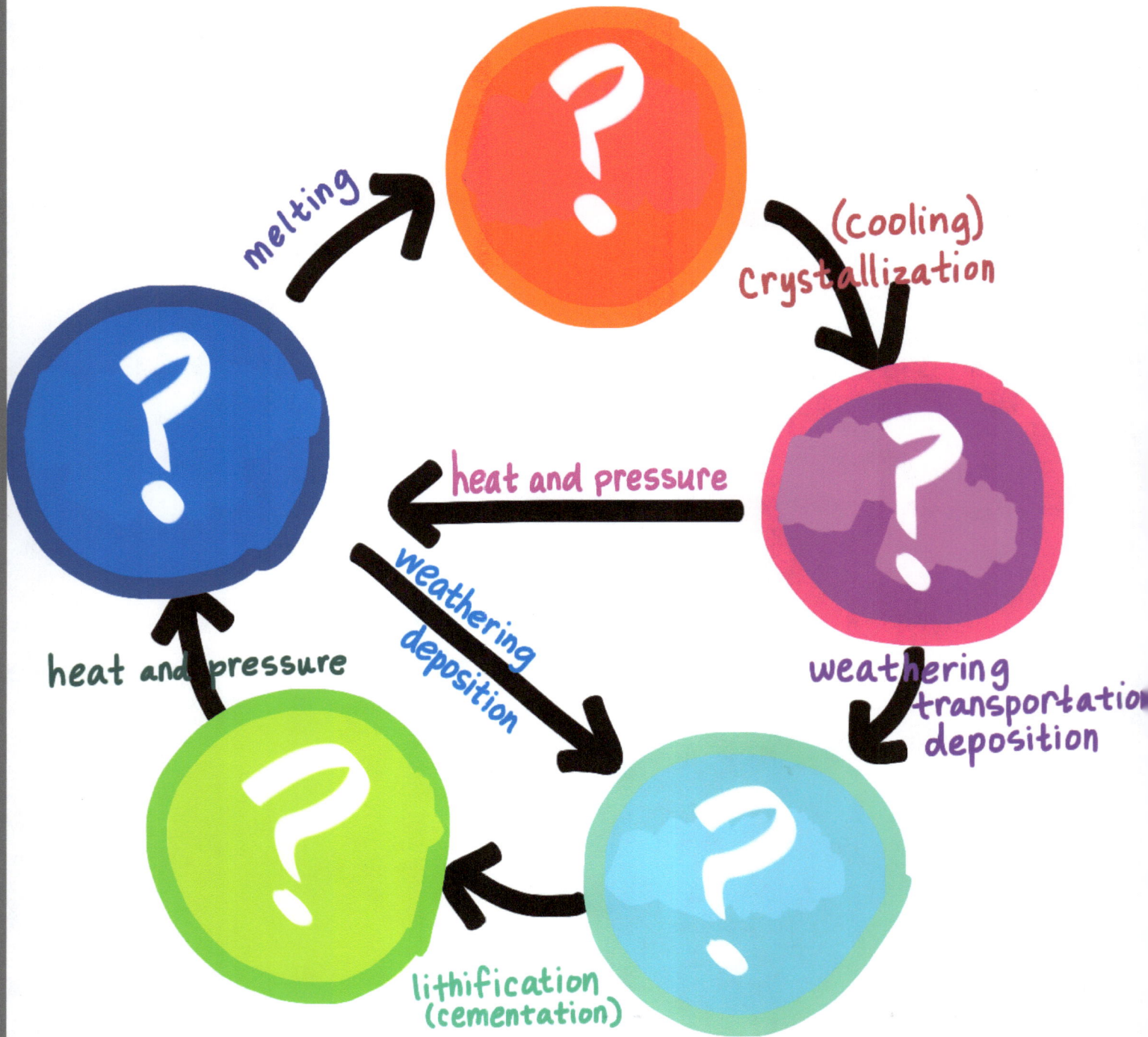

melting

(cooling)
Crystallization

heat and pressure

weathering
deposition

heat and pressure

weathering
transportation
deposition

lithification
(cementation)

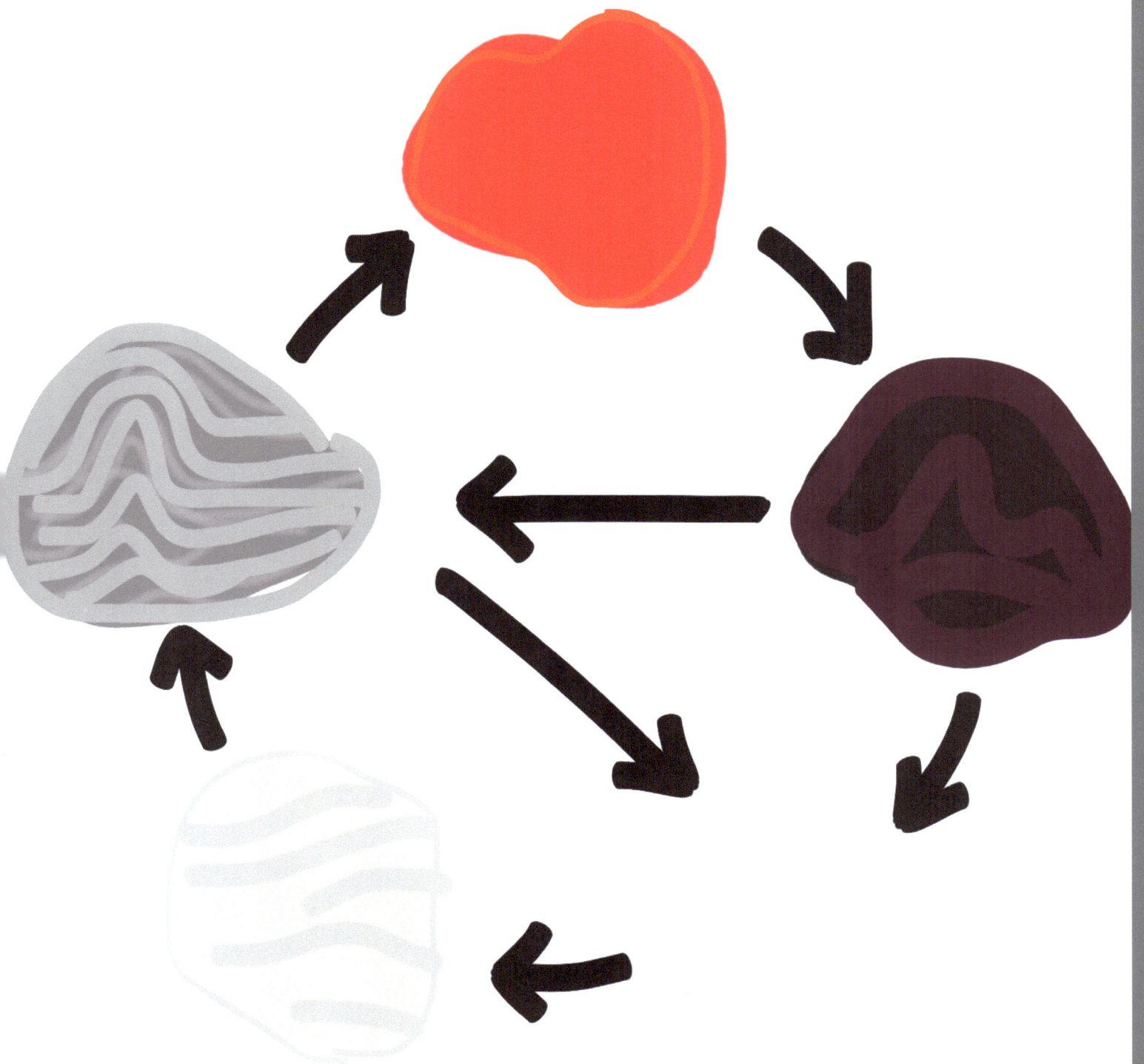

Well done!

Cloud
(KLOUD)

Precipitation
(PRE-SIP-EH-TA-SHUN)

Water
Cycle

Percolation
(PUR-KO-LA-SHUN)

Aquifer
(AW-KWEH-FUR)

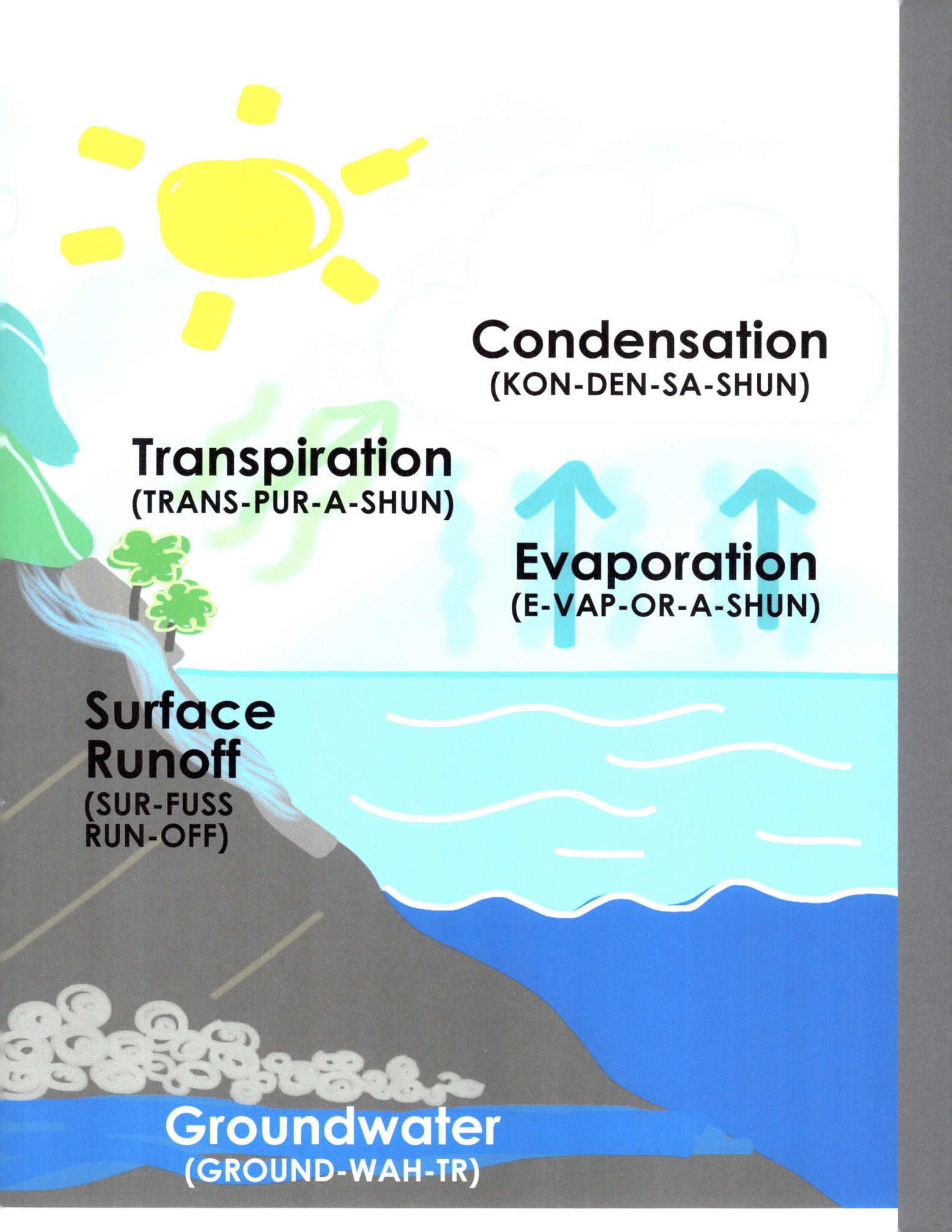

What does it feel like when the sun is out during summer?

Do you sweat?

Did you know that plants sweat?

Plants lose water out of their leaves, just like we lose water out of our pores.

Transpiration is plants losing water from their leaves into the atmosphere as water vapor.

Evaporation occurs when the sun heats oceans, lakes, rivers (any body of water) and turns water into vapor.

Water vapor goes into the atmosphere and gets very cold, causing it to become liquid water again in the form of clouds. This is the process of condensation.

(Condensation also occurs when you have a cold drink outside on a hot day)

When the cloud gets too heavy with water, precipitation occurs.

Precipitation is when water falls back to the surface of the earth as rain, sleet or snow.

After <u>precipitation</u> occurs, water enters the ground and goes through <u>percolation</u>.

<u>Percolation</u> is the movement of water through the ground (layers and layers of soil and rocks).

<u>Groundwater</u> contained in permeable (PUR - MEE - UH - BL) rock is called an <u>aquifer</u>.

Permeable means liquids and gases can move through it.

Aquifers provide water for us to drink. **Precipitation** and **surface runoff** goes into **aquifers**.

Chemicals and toxins can contaminate (KUN-TAM-IN-ATE) **aquifers**. We take care of the earth and our water supply by **NOT littering or polluting** on the surface where water is collected.

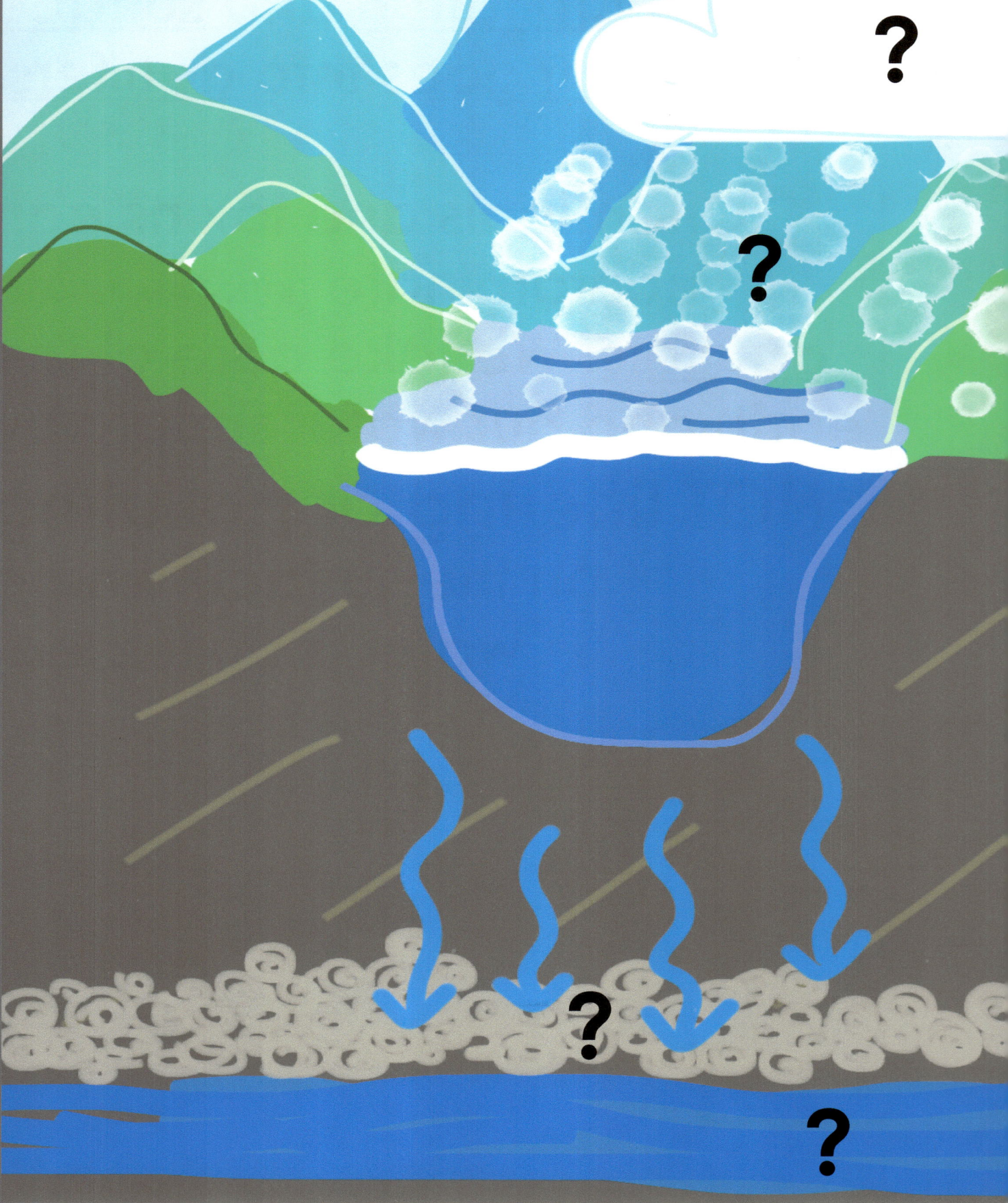

Good Job!

Geology is the study of how the earth was formed, and what it is made of.

Earth is composed of four main layers: the inner core, the outer core, the mantle and the crust. The inner core can reach 5,000 degrees celsius!

The four spheres of the earth interact with each other: the hydrosphere, atmosphere, biosphere and lithosphere.

Tectonic plates are part of the earth's crust, and therefore part of the lithosphere.

Volcanoes form where tectonic plates meet.

When two tectonic plates converge, one plate goes below the other, creating earthquakes. The pressure causes magma to erupt from volcanoes. Ash and lava layer on top of each other, gradually making a larger and larger volcano.

There are many landforms on the earth that create a varied and beautiful landscape. How many landforms can you remember?

The rock cycle shows how the three main types of rock (igneous, sedimentary and metamorphic) change into different forms through heat, pressure, weathering, erosion.

Water is <u>evaporated</u> by heat from the sun, pulling water vapor into the atmosphere. Plants release water into the atmosphere through <u>transpiration</u>. As water vapor enters the atmosphere, it becomes cold, and the vapor becomes water again forming clouds. This is <u>condensation</u>. When the clouds become full of water, <u>precipitation</u> occurs, and water is returned to the surface by rain, sleet or snow.

Our drinking water comes from the surface of the earth! Protect our <u>aquifers</u> and water supply by not littering or polluting.

Congratulations, YOU ARE A GEOLOGIST!

Geology	Mountain	Igneous
Geologist	Lake	Sediment
Inner core	Butte	Sedimentary
Outer core	Plateau	Metamorphic
Mantle	Mesa	Solidifiying
Crust	Strait	Evaporation
Magma	Cape	Transpiration
Continental	Islands	Condensation
Oceanic	Harbor	Cloud
Hydropshere	Hills	Precipitation
Atmosphere	Glacier	Surface Runoff
Biosphere	Waterfall	Percolation
Lithosphere	River	Permeable
Tectonic plates	Delta	Aquifer
Converge	Beach	Contaminate
Volcano	Gulf	Groundwater
Lava	Ocean	
Conduit		
Vent		
Crater		

...and most important,
protect your water supply!
NO LITTER, NO POLLUTION

Well done!

You are a
Anatomy,
Cardiology,
Botany,
Marine Biology
&
Geology
EXPERT!

www.ingramcontent.com/pod-product-compliance
Lightning Source LLC
Chambersburg PA
CBHW051600190326
41458CB00029B/6492